FACES ON THE FRONTIER
FLORIDA SURVEYORS AND DEVELOPERS IN THE 19TH CENTURY

JOE KNETSCH

THE FLORIDA HISTORICAL SOCIETY PRESS
COCOA, FL
2006

Faces on the Frontier: Florida Surveyors and Developers in the 19th Century
Joe Knetsch
Copyright 2006

Published by the Florida Historical Society Press

All rights reserved under International and Pan-American Copyright Conventions. No part of this book may be reproduced in any form or by any means, electronic or mechanical, including photocopying, recording, or by any information storage and retrieval system, without permission in writing from the publisher, except by a reviewer who may quote brief passages in a review.

ISBN: 1-886104-24-7

The Florida Historical Society Press
435 Brevard Avenue
Cocoa, FL 32922
www.fhspress.org

P•R•E•S•S

TO
LINDA

TABLE OF CONTENTS

Preface i

Introduction: iii

Part I. Surveyors General
1. A Brief Life of Robert Butler: Surveyor General of Florida 1
2. Benjamin A. Putnam 9
3. A Finder of Many Paths: John Westcott and
   the Internal Development of Florida 27
4. Untiring, Faithful and Efficient:
   The Life of Francis Littleberry Dancy 47

Part II. The Surveyors
5. Colonel Sam Reid: The Founding of the Manatee Colony 63
6. A Surveyor's Life: John Jackson in South Florida 77
7. D. A. Spaulding: A Special Man with Exceptional Talent 93
8. Surveys and Surveyors of Southwestern Florida 99
9. Marcellus Stearns' Report on South Florida in 1872 109
10. Benjamin Clements 119
11. R. W. B. Hodgson and the Origins of the Whitner-Orr Line 131
12. Charles H. Goldsborough 137

Part III. The Developers
13. Forging the Florida Frontier: The Life and Career of
    Captain Sam E. Hope 149
14. Impossibilities Not Required: The Surveying Career of
    Albert W. Gilchrist 171
15. Hamilton Disston and the Development of Florida 185

Conclusion 207

Index 213

FACES ON THE FRONTIER
FLORIDA SURVEYORS AND DEVELOPERS IN THE 19TH CENTURY

JOE KNETSCH

Township Layout — Surveyor's Notes

PREFACE

Looking out over the dreamy landscape of the deep Everglades with its cypress islands, waving saw-grass and flights of exotic birds one can only wonder why this portion of Florida was not surveyed. The answer is simple, it costs too much to meander those islands and wade through the swamps and saw-grass. It did not cost as much to survey the remainder of Florida, however, and the task was nearly as difficult as measuring the Everglades. Florida had numerous swamps, rivers, streams, lakes, ponds, bayous, etc. and they all were expensive to survey. But the settlers wanted the land, the government needed the money and Florida got surveyed in spite of itself.

The essays collected in the following chapters explain some of the major problems of surveying this wonderful state. Through the eyes of the surveyors and their bureaucratic colleagues in the General Land Office or Surveyors General Office we see a different Florida than that presented in guide books or most academic tomes. Written over a period spanning some fifteen years or more these essays represent a portion of the research done in search of the reality of frontier surveying. Few of the men represented here have ever been mentioned in the textbooks or other writings on Florida and that is part of the joy in researching their lives and presenting them in print. Many of these men were the actual leaders in their communities and were politically connected to those who have made the "great man" list of Florida leaders. All faced some incredible odds in accomplishing their tasks and each persevered through them to successfully give us our land boundaries as they stand to this day. It has been a pleasure researching and writing the stories of these pioneers.

Most of the essays that follow have seen the light of day in print elsewhere. Many were in one form or another published in *Florida Surveyor*. The *Sunland Tribune*, the magazine of the Tampa Historical Society, was kind enough to publish at least four of the articles concerning the surveyors and developers of that area. *El Escribano*, the magazine of the St. Augustine Historical Society, published the piece on Benjamin Putnam after I had presented it orally to the Putnam County Historical Society, where I also presented the chapter on F. L. Dancy. The Florida Historical Society, through its *Journeys* magazine, published the introductory chapter and the piece on John Westcott was published in *Florida Pathfinders*, an edited volume by Lewis N. Wynne and James J. Horgan. The author is grateful for the guidance and editorial experience of those who worked so hard on those publications and who made the essays more readable and understandable to their readers. The responsibility for the content of these essays and articles still belongs to this author. Although each essay has been edited before and often reviewed by my colleagues at the Department of Environmental Protection, Division of State Lands, the content reflects only the opinion and research of this author and is not intended as the official position of the Department or its staff.

INTRODUCTION

Few textbooks tell the story of the men who brought order to the frontiers of America. By order we mean to measure the land into sections, townships, ranges, blocks and lots. Before anyone can obtain clear title to the land, it must first be surveyed and given distinction so one person can tell his/her property from another's. The men who performed this task were often the very first non-Native Americans to see the land in its natural state. The descriptions they have left us, in the form of field notes and plat maps, are the closest we can get to a true picture of the physical frontier. Yet, their story is almost forgotten in the textbooks, official histories and biographies of those who have benefited most from their work, all of us.

It is not difficult to imagine some of the territory of Florida in its primitive state, one has only to look at the wild beauty of the Everglades National Park. But most of Florida was not everglades, but a widely varied land of rolling pine covered hills, with "blackjack" oaks and other hardwood trees sprinkled among them. Some of the land was covered with prairie-like grasses that frequently burned in the lightening charged air of middle and southern Florida. Still other portions of the territory were covered with cypress swamp and thick "baygalls", as the wet, entangled titi and wax myrtle scrub of the low lying areas were often called by the early settlers. All of these types of environments are described in the field notes of the U. S. Deputy Surveyors and all of them were surveyed by these very same men.

To survey the land means to mark it out in an organized pattern, such as sections, townships and ranges. Florida, like all public lands states, is divided up into such a pattern. First, a "prime meridian" (north to south) line is marked out. In Florida this line runs from the Gulf of Mexico, near St. Marks to the Georgia border and runs through the capitol of Tallahassee following the line of Meridian Street. A second line is run, called the basis parallel line, and it goes from Tallahassee east to near the southern end of Little Talbot Island, in Duval County, and westerly from Tallahassee to the Perdido River, the boundary between Florida and Alabama. Because of the great amount of water in Choctawhatchee Bay and St. Andrews Bay, there is a "jog" in the line running west. The basis parallel and the prime meridian lines intersect in Tallahassee (the spot is marked by a monument).

From this intersection point, all surveys of Florida begin. Lines running parallel to the basis parallel line are referred to as "township lines" and those paralleling the prime meridian line are called "range lines". The unit called a township is a square of 6 miles and this square is divided into 36 sections. This system was developed, in its early stages, by Thomas Jefferson and refined by the Land Ordinance of 1785. It is officially called the "rectangular system" of surveying.

Aside from the need to have an organized system of land measurement, the surveys served another very important function. They told potential settlers exactly what the land was like in a given township. Just how important is this information? Say a settler from New England, used to growing corn, potatoes and pumpkins wants to come to Florida, where all three crops can grow, but only in specific areas. The New Englander would not want to move his family and belongings to an area of high, sandy hills where these crops will probably fail and leave the family without food. Therefore, by consulting the field notes at the local land office, these potential settlers will find the area of Florida where the crops they wish to grow will prosper.

To plot out this land and give the information to the potential settler was the major job of the surveyors. This meant, as noted before, that these men were often the first non-Native Americans to see the lands of Florida in their wild state. Spanish and English explorers did cover much of the state, however, they did not leave enough accurate information behind to be useful and the settlements they founded, like St. Augustine and Pensacola, were small and often isolated from the mainstream of colonial activity. This left the task of accurate description to the United States Deputy Surveyors, under the leadership of the Surveyor General for Florida. The job was not easy and full of danger.

When Florida was acquired from Spain, in 1821, it was not well mapped and virtually unknown in the interior. The first Surveyor General, a former ward of Andrew Jackson, Robert Butler, somewhat miscalculated the time needed to survey the first lands to be opened for settlement. He and his deputies were delayed by heavy rains, mixed and confusing instructions from the General Land Office in Washington D. C. and the difficult nature of the lands in northern Florida. Additionally, the surveying season, because of the lack of medical knowledge, was limited to four to six months of the year, or only during the dry months. It was believed that dangerous and fatal "miasma" rose from the lands during the rainy season and no one would work during this period. This was also believed by the United States Army, which campaigned against the Seminoles, Miccosukees,

Creeks and others only during the "healthy" season, or dry time. Thus, when Colonel Butler, as he was known to most, said he would have his men finish 130 townships in one year, he was off by a large margin. Not until the surveying season of 1824-26 did the speed of surveying pick up and larger quantities of lands sold to the public.

Surveying in the early years of Florida was very difficult. The surveyors and their crews had to face numerous hardships. Many men died from disease during the first few years, most notably the crews who surveyed in the area of Pensacola and the Escambia River basin, who contracted the dreaded "yellow fever" in the late 1820s and early 1830s. One surveyor, Benjamin Clements, an old friend of and soldier under Andrew Jackson, lost his son, Hosea, and one other crew member to the fever and four others had to be hospitalized for an extended period. Malaria, dysentery and various forms of intestinal diseases afflicted the early crews of the surveyors. The terrain, too, was often rugged and a severe test of the stamina of these pioneer surveyors. Swamps, bogs and Florida's numerous water bodies provided difficult challenges to these frontiersmen. Few were prepared for tackling the twisted and knotted cypress swamps, particularly the great "Green Swamp" of central Florida, headwaters of four major river systems. The tidal marshes, also, gave surveyors a great deal of difficulty. Tramping through the swamps in Florida's cool winter months was not a pleasurable task.

The great variety of pests, known to all as insects, also plagued the surveyors, many of whom had never seen mosquitoes so large or sand fleas so hungry as those in Florida. Snakebites were common for the surveyors and, although seldom fatal to most victims, they did make many very ill for three to five days. As with any profession requiring physical labor, accidents happened that laid up surveyors or members of their crews for many days. Simple exhaustion often delayed surveys, because the surveyors had to work in all types of weather to cut (mark) lines, take measurements and literally make roads (paths) to get to their work. Poor planning often delayed the work for weeks at a time. One should never forget that the surveyor in the field had to provide the food and tools for the crew, plan the work so as to keep everyone productive as long as possible, buy the horses or mules and the wagons to haul the materials and food, pay for the instruments used to measure the land and, finally, pay everyone of the workers on his crew. This meant that the surveyors had to have money or people willing to pay the expenses (bondsmen) and be willing to wait, sometimes as much as two years Deputy Surveyor was ever killed in the line of duty by Florida's Indian population. This does not mean that confrontation did not occur. John Jackson, a deputy surveyor from Tampa, had some memorable encounters with Indians, including one where his campman/cook, an African-American, was told twice to leave the territory under pain of death and the poor man was, in Jackson's words, "nearly frightened out of his wits." On another occasion, the Indians "fired the prairie" behind the surveyors forcing them to leave the vicinity. Upon waking up one early morning in 1855, surveyor W. S. Harris found his horses missing. He spent four days trying to find them and when he did eventually locate his horses, they were hobbled, Indian style, in the Kissimmee prairie. Yet, even when the surveyors were being used by the government to edge the Indians out of Florida by surveying the "neutral ground" of 1842, the Seminoles and Miccosukees did not attack the crews, even though they had every opportunity to do so. The surveyors were not well armed, the U. S. Army was and for the Indian peoples, that was the major difference.

Of all the problems faced by the surveyors on the Florida frontier, loneliness was one of the worst. Thoughts of loved ones, family and friends often filled the letters of the surveyors. It could be weeks or months before a surveyor in the field would or could receive or send a letter to his family and friends back home. Since many of the surveyors were leaders of their respective communities, being out of touch with what was going on added to the pangs of loneliness. In the late 1840s, surveyor Henry Wells, noted for his dry humor, wrote to the register of state lands, John Beard, a political and social friend: "I wish you would file away the list among my papers in the land office provided you are still the incumbent, if not, please hand or send it to that functionary...I have not heard from the white settlements for a long time. Anything from any body would be truly acceptable." The isolation a surveyor felt in the wilds of the territory of Florida was strong and is expressed often in their correspondence with other officials.

As leaders of their local communities, the Florida surveyors stand out as a group. The examples of their leadership abound. Dr. John Westcott, for example, served in two constitutional conventions, was the president of a railroad, became noted as the "Father of the Intracoastal Canal" and served in the State's first legislative session, where he helped to shape today's educational system. Captain Samuel E. Hope was the founder of the community of Anclote, which later became part of today's Tarpon Springs. He also served in three legislatures and two constitutional conventions. Captain Hope served in the Third Seminole War and was elected captain of his unit during the Civil War. Major Romeo Lewis had the distinction of serving as sheriff of both Leon and Jackson Counties during his lifetime. Surveyor General Benjamin Putnam was one of the most respected military men and judges during Florida's years as a territory and early in its statehood. Putnam County is named for this distinguished gentleman. Finally, Governor Albert W. Gilchrist, of Punta Gorda, began his professional life as a surveyor for the railroad and for the United States. It should also be noted that these men followed in a long tradition of surveyors as leaders of their communities, states and the nation. George Washington, Thomas Jefferson and Abraham Lincoln were all surveyors at some point in their lives.

Frontier surveyors were generally well-educated individuals. Their training, as a group, stands out in the very accuracy of their work. True, there were some who did not fulfill their functions properly, however, the majority of these pioneer surveyors showed that a good foundation in mathematical education paid dividends to the society that followed their lines. To this day, the survey lines of a Henry Washington, an A. M. Randolph or a Benjamin Whitner Jr., are so good, given the crudeness of their instruments, that modern surveyors can often follow them to the exact point of departure and ending with little difficulty. Each of these men, and their numerous colleagues, had good educations and a firm understanding of the principles of surveying.

Henry Washington was the nephew of General William Washington of Virginia and later South Carolina, and related, though vaguely, to that other surveying Washington, George. His immediate background is unknown, however, he was reared in the traditions of the genteel South and this often included some military training, which, in the early years of this nation, meant a good grounding in mathematics. He gained valuable experience in his early career in the agricultural fields of what is today's Mississippi and Alabama, where he worked out of the Washington, Mississippi, office of Surveyor General John Coffee. His work was highly thought of by his colleagues and nine of these gentlemen signed a joint letter of endorsement when he applied for his first surveying position in Florida. One letter of endorsement, from Levin Wailes, an early leader in Mississippi, noted the character of Washington, "Without disparaging other Surveyors it is but an Act of justice due to Mr. Washington to state that his returns have been among those which have merited the highest approbation for their accuracy & perspecuity. Indeed he never seemed satisfied with himself but when he has made practice approach theory the nearest of which it is susceptible."

Henry Washington strived for perfection in a profession where it was, and still is, difficult. He worked in the wilderness areas of Mississippi, Alabama, Florida, Louisiana and California with imperfect instruments and tremendous hazards. He braved the heat of the Florida and Louisiana swamps and the cold of the California mountains to plot lines that became the basis for the property of millions of people. And in Florida, he succeeded in an arena fraught with the difficulties of political actions, especially in the survey of the "Great Arredondo Grant." Nearly every surveyor who has ever followed the landlines he established, however, recognizes his achievements. As the historian of the Bureau of Land Management put it in *Surveys and Surveyors of the Public Domain: 1785-1975*, "Everywhere that surveyors have retraced his lines, whether in Florida swamp or California desert, their conclusions have been unanimous: Henry Washington was one of the best [surveyors]."

Two of Washington's colleagues in the field were medical doctors, Arthur Morey Randolph and John Westcott. A. M. Randolph came to Florida at a young age and was sent off to school at the University of Pennsylvania for training in the medical profession, which, in those days, involved experimenting with measurements of the human anatomy to help diagnose diseases of the body. Arthur, according to family lore, had studied some engineering prior to his attending the famed medical school. Upon his return to Florida, he went into partnership with his brother, the well-known physician James Randolph. Tallahassee, which was not big enough to support two physicians, failed to offer Arthur the opportunities he needed to support himself and his bride, the daughter of Governor William Pope Duval. He, therefore, took up surveying and began a career that included surveying millions of acres of the State of Florida. Indeed, probably no other man surveyed as much land or for so many reasons than A. M. Randolph. His success as a surveyor soon allowed him to set up his own plantation on the outskirts of Tallahassee on the old mission site of San Luis where he built a steam-driven cotton gin and a hydraulic ram to pump water for the plantation's fields and gardens. Like Henry Washington, Randolph also surveyed many of the remaining Spanish land grants in Florida, a difficult task because many were never surveyed by the Spanish. In these special surveys, A. M. Randolph had to convert many of the Spanish measurements into their contemporary American equivalents, a task which presented many difficulties because of the many variations in the Spanish measurements. His surveys of these grants, with one notable exception, remain the standard in many areas is testimony to his attention to mathematical detail in the conversions and the actual measurements.

But Arthur Randolph was not only a surveyor of grants and township lines. His surveying experience carried him into the realm of numerous specialized surveys and unique requests upon his talents and judgment. In particular, A. M. Randolph, with Henry Wells, another good Florida surveyor, was chosen by the governor to become the first selector of lands under the Swamp and Overflowed Land Act of 1850, the most important land act passed concerning Florida and the largest single grant of land to any state in our nation's history. This act brought over 20,000,000 acres of land under state control. The first selections were crucial in determining the future growth of the state. He also was one of the team of surveyors chosen to select the first "Internal Improvement" lands, by which the state of Florida received 500,000 acres of land, the sale of which was to promote railroads and canals. Randolph's career involved him in the Seminary lands selection and surveying lands permitted under the Armed Occupation Act of 1842. Because of his established reputation for accuracy and tenacity, A. M. Randolph was constantly trusted with the difficult jobs, which sometimes put him at the wrong end of a gun barrel, such as when the citizens of Alligator (today's Lake City) refused to let him survey the "Little Arredondo Grant" for fear of losing their homesteads. In this famous case, discretion was the better part of valor and Randolph did not survey the grant, thus saving the lives of his crew and, maybe, many more individuals as well. In each of these special survey cases, Randolph's judgment was heavily relied upon to bring each to its proper close. It is always to be remembered in studying early surveying, that much was left to the discretion of the early surveyors. If their judgment was in any way faulty or erratic, their work could not and cannot be relied upon. The career of Arthur M. Randolph demonstrates that his judgment was sound in almost every instance, a true measure of his character and reliability.

The other physician in the ranks of Florida's early surveyors was Dr. John Westcott, the younger brother of Senator James D. Westcott. Like his colleague, Arthur Randolph, Westcott's hometown, Madison, was too small for two doctors. Dr. Westcott, after service in the Second Seminole War, began to cast around for another career and decided to use his natural inclination for mathematics in the profession of surveying. Westcott is known among today's surveyors for his meticulous work in the field. His sense of duty to the profession also made him one of the very few surveyors in Florida history to follow all directions given in the Manual of Instructions as closely as the terrain permitted. If the directions called for him to meander a twenty-five-acre lake in the middle of a section and tie his meander posts into the nearest quarter-quarter post, he did so to the best of his ability. When he became Surveyor General of Florida in late 1853, he demanded the same of those surveyors who worked under him. To make sure that his and the General Land Office instructions were being carried out in the field, he frequently went on personal field inspections of the work in progress and, when this was not possible, he hired competent surveyors to check the work of their colleagues, thus insuring the greatest possible accuracy in any work approved during his tenure.

John Westcott was more than a Deputy Surveyor or, even, a Surveyor General. He was an inventor and innovator of the highest order. In all cases, his mathematical instincts and education played an important role. In Madison, he built one of the first steam sawmills to operate in that region. In the early 1870s he worked on an invention that was displayed for the world to see in the Centennial Exposition, a "saddle-bag railroad" line. The main feature of this invention was a single track with a shoe-like devise propelled along in front of an engine driven by wheel-traction with the shoe inside of the track instead of straddling it like a railway car's wheels. This, as some will recognize, is a proto-type of today's monorail. (He even established a company to test this out in Florida, but he attempted to build it on the ground between Orange Lake and the Ocklawaha River where the loose sand created too much friction for it to work.) However, even though his inventions failed to bring him fame or fortune, he did begin two ventures that paid huge dividends for the citizens of Florida. The first was the organization of the St. Johns Railroad, from Tocoi to St. Augustine, the first successful attempt to link the St. Johns River to the Ancient City. This line later, after the War Between the States, became part of the Flagler system of railroad lines. The second venture was the organization of the Florida Coast Line Canal and Transportation Company, the forerunner to the Intracoastal Canal. In each case, the initial lines of the routes was surveyed and plotted by Westcott himself. He was nearly eighty years of age when he ran the last of the route for the coastline canal.

Westcott's remarkable career also included a stint in the first Florida legislature, where he pushed his now famous plan for education in Florida. In this plan he stated a basic truism for all to remember, "Education is to the Republican body politic, what vital air is to the natural body; necessary to its very existence, without which it would sicken, droop, and die." He advocated a system of public libraries too which would be open to all in society and contain books, "chiefly works on morals, natural philosophy (mathematics), Natural history, Geography, Agriculture, Astronomy, History, and Biology, Chemistry & Physiology, and

Political Economy." His emphasis, as can be seen from this list of general works, was placed upon those sciences that are rooted in the study of mathematics. Westcott, like all surveyors, realized the very essential nature of mathematics to the fullness of human life and labor.

Not every surveyor, however, was to be taken so seriously as Washington, Randolph or Westcott. Not that these gentlemen did not display appropriate humor in the proper circumstances, but there was one fellow surveyor among them who, for sheer volume, audacity and bombast stands out in the records of Florida surveying. His name was John Irwin. In the letters of "Bombastic John," my nickname for this readily readable gentleman, are found some of the most unique phraseology found in the history of surveying. "In his own words" is the best way to appreciate the linguistic ramblings of this peripatetic purveyor of verbosity:

> Sir: It is with mingled emotions of pleasure and pain that I respond to your brief but very welcome letter of the 12th ult. for I am mortified and depressed at my egregious oversight in not signing the hasty letter I wrote you on the 16th of last Oct. At the time of writing it, the man who was to take it to Marianna was impatiently waiting for it, while at the same time the men were importuning me, and clamouring for money, shoes and clothes, of which they stood extremely in want: for they were not only in a state of nudity, as their garments of many colours, and many patches, were almost all frittered away among the tie-tie scrubs, cypress swamps, bushes, and briars, of the Choctawhatchee Peninsula: but they also could display more scars and scratches than any of the war worn soldiers of the renown Napoleon. This no doubt contributed to confuse me, and commit the unconscious error; besides, after I had rehearsed the letter I concluded not to send it on account of its many faults. And yes, I had no other paper to copy it, nor any other opportunity of sending it to the Post Office and for the necessaries we stood in need of at the same time. But perhaps the best palliation I can offer is to plead guilty to the <u>Irish</u> blunder and to ask forgiveness of an indulgent friend; and to promise with Divine assistance to be guiltless of the like in the future.
>
> As I always sympathized in your trials, and rejoiced in your triumphs ever since I had had the honour and happiness of your first acquaintance it would seem almost supererogatory to say that I experienced similar sensations at the perusal of the statement of your health in your last letter. For the sensitive mind of the grateful recipient like every other thing very elastic when relieved of its burden rebounds with

increased vigour even beyond its wonted equanimity at the prosperity and welfare of a good friend and benefactor.

And this letter goes on for an additional three pages in similar fashion. On the last page, however, indicates, albeit rather fawningly, that even in the wilds of the Choctawhatchee swamps he could not forget that his basic task was a mathematical one:

> I flatter myself that you will be pleased with the methods I pursue in ascertaining the width of the creeks, rivers, &c., Allow me to give you 1 or 2 specimens briefly; in running West I come to the bank of the river, I retrograde 43 lks through choice, and sight to an object due West; then I again take S. 30 W. as another sight, and go on this course until such time as the object on the opposite bank bears N. 30 W. which is the distance of 5.97 chs. hence I have per 32nd 1st b. Euclid, 3 angles and 1 side of an equilateral triangle viz. 60 & 5.97 chs. therefore 5.97 - .43 = 5.54 chs. the exact width of the river in this place.

His second example, not to be given here, involved the angles and one side of an isosceles triangle. In the deep woods of west Florida, a surveyor spends his evening by the campfire writing the surveyor general of the harsh conditions of work and still includes his basic math lesson for the day, as if the surveyor general needed one so late in the day.

Bombastic John Irwin is fun to read and entertaining to describe. However, the lesson here is that, even though it sounds silly to the modern ear to hear of a surveyor describing basic mathematics to the surveyor general, the need to reassure himself of these basic principles of mathematics was important to John Irwin. He felt the need to let his superior know that he knew exactly what he was doing in running his lines and thus give his superior confidence in his work product. What better way to express this need and prove the point than to run through the lesson, citing Euclid's theorem, book and verse?

In all of the above examples of early Florida surveyors, the theme has been their reliance on mathematics, education and self-motivation. Each man was meticulous in his work and accurate for his day and time. The hardships, indicated here only by Irwin's letter, danger, and harshness of the climate are only to be added to give us an idea of the immense struggle each undertook to give us the basic lines with which to establish our property and perfect our inheritance.

In establishing the boundaries of private and public property throughout the state, the surveyors of Florida's frontier performed and invaluable and necessary function. All land titles in the state depend upon the surveys of these early pioneers, which means that all titles to land in Florida can be traced back to the original work performed by these remarkable, yet often forgotten, men. To bring what Europeans called civilization to the frontier required that the land be parceled out to individual owners who would make the land productive of crops needed by everyone to live. The rectangular surveying system, founded upon Jefferson's work and developed by the United States government, made sure that the land was correctly surveyed and marked out for the benefit of individual owners.

Without the surveys, performed under hazardous conditions and in the unmarked swamps of early Florida, our world would be much different and more confusing place in which to live.

Joe Knetsch, PhD
Tallahassee
May 1, 2006

CHAPTER 1

A BRIEF LIFE OF ROBERT BUTLER: SURVEYOR GENERAL OF FLORIDA

On February 22, 1824, Andrew Jackson wrote to John Coffee the following, "I found shortly after my arival here that the surveyor Genls office for the Floridas was intended for Colo Preston late governor of Virginia, he had been appointed commissioner of land claims, held it nearly two years, recd the emoluments, and never went there. I had, in justice to Colo Butler, to interpose his claim, and bring to Mr Monroes recollection his promise to do something for Butler, and I have obtained his promise that he shall be provided for in Florida, and Mr Preston provided for otherwise."[1] Thus, through the direct intervention of Andrew Jackson, Robert Butler became the first Surveyor General of Florida in 1824. This office carried with it very heavy responsibilities, including setting out the territorial capitol, hiring the first surveyors and establishing the integrity of the surveys offered for acceptance. In performing these duties over an extended period of time from 1824 to 1848, Robert Butler performed a service for the State of Florida that should never be forgotten.

Butler began life on December 29, 1786, the eldest son of Colonel Thomas Butler, a Pennsylvania born Revolutionary War veteran and career officer in the

---

[1] The following sketch of the life of Robert Butler was presented on behalf of the Northwest Florida Chapter's nomination of the Butler grave site for the National Register of Historic Places. Its purpose was to quickly show the importance of Florida's first Surveyor General to the history of the Territory and the Nation. His life needs little introduction to the professional land surveyors of Florida because each follows, to some degree, the dictates, orders or instructions Butler sent to the pioneer surveyors of Florida's vast frontier. In the daily life of most surveyors, the name of Robert Butler appears on plats, instructions or some other historic document that sets the pattern for public land surveys in our state. It is hoped that this brief biography will add a bit more depth to our understanding of this cautious and steady leader of Florida's first surveys.
The Correspondence of Andrew Jackson, Volume III, 229.

United States Army. The life of a career officer led Thomas Butler through many moves and family displacements. One of the moves, however, put him in Davidson County, Tennessee, and a near neighbor to Andrew Jackson, with whom he became close friends. Upon Thomas Butler's tragic death in 1805 from the Yellow Fever epidemic in New Orleans, at the request of the father, Andrew Jackson extended a "father's guidance" over the career of young Robert A. Butler. At this point in his life, the young Butler was over six feet in height and noted for his fine physical appearance. Being reared in a military family, young Robert A. Butler turned to the military for his career, although the exact nature of his training is unclear.[2]

Being the eldest son also gave young Butler much of the estate that his father had accumulated, including plantation lands near to Jackson's. The nearness to Jackson and the society of Nashville gave Butler some advantages in social learning and economic advancement. One of the main advantages was the love of horse racing, the favorite sport of the planter class. Also a lover of this sport was Rachel Jackson, the General's wife, and her favorite niece, Rachel Hays. Miss Hays was one of the more attractive and sought after young ladies of her day and Robert Butler became the victorious suitor. They were married at the "Hermitage" on August 29, 1808. The marriage not only made Butler a "relative" of Andrew Jackson, but also of the Hays and Donelson families, two of the more important families in frontier Tennessee. The marriage also brought other economic advantages that allowed Butler to advance in the amount of land and slaves owned. Robert Butler was well on the way to becoming one of the more successful planters in western Tennessee.

At this point in his life, the United States went to war with Great Britain in the War of 1812. In recognition of his previous military training and political contacts, he began the conflict with the rank of captain. According to his "memoir", published in the Nashville *Union*, in 1849, Butler served in some of the more important campaigns in the western theater of the war. He served in the investment of Camp Meigs under General Clay Green, of Kentucky, in early 1813. He accompanied General, later president, William Henry Harrison on his campaign into Upper Canada and fought at the Battle of the Thames. He was present at the re-occupation of Detroit, serving as the Adjutant-General for the Eighth Military Department. Presumably, he was promoted to the rank of major with this assignment, for when he is transferred to the Seventh Military Department, then under

---

[2]Mary L. Davis, "Robert Butler: An American Pioneer," 1939.

Andrew Jackson, he is listed as "Colonel" Robert Butler. For Butler, one of his most important services was the service on a board of officers that was "mainly instrumental in saving the troops from famine," in Detroit, the lakes being frozen over in that year.[3]

His transfer led him into the thick of the fighting verses the Creek Indians in southern Alabama. Here he raised troops to assist in the fighting under General John Coffee, later Surveyor General of the lands south of Tennessee. The object was to chastise the commander at Pensacola, who was correctly suspected of providing the Creeks with weapons. Jackson's force soon penetrated Florida and proceeded to Pensacola, where, by rapid movements, the General essentially surprised the enemy and with quick charges took the out-lying batteries. After entering Pensacola and dispatching the Indian faction, Jackson's men, including Butler, returned to Mobile, reenforced its defenses and headed toward New Orleans, the suspected target of British operations.[4]

The transfer to the Southern District, as Jackson called it, brought him into contact with many of his former friends and future associates, including Richard Keith Call, two-time governor of Territorial Florida, who frequently shared mess with him. Call also recalled sharing the privations of campaigning and his tent with "the friend of my youth."[5] He was soon assigned recruiting duty in western Tennessee. Butler arrived back at headquarters in time to actively participate in the famed Battle of New Orleans. According to Call, he was heavily engaged in the fighting on January 8, 1815, when General Packenham and many others fell to American arms. For his part in the battle, he was one of three officers chosen to accept the surrender from the British commander at the close of the battle.[6] After this service, he immediately resumed his recruiting duties, assisting General Coffee in raising a force to fight the "Red Stick" faction of the Creeks.[7]

Shortly after the triumph at New Orleans, Jackson and his staff returned to Tennessee. Although Butler had been assigned to assist Coffee, he soon was back at his plantation attempting to get his affairs in order. In December of 1817, Jackson received orders to move his men out and drive the Creeks and Seminoles out of

---

[3] Davis, 51.
[4] *Ibid*
[5] J. Roy Crowther, *The Grand Lodge of Florida Free and Accepted Masons History, 1830-1989* (Jacksonville: Drummond Press, 1990), 10.
[6] Donald E. Merkel, *Colonel Butler and the Public Land Survey of Florida (1824-1849)* (Tallahassee: Florida Department of Transportation, 1974), 2.
[7] Davis, 52.

American territory and then out of Florida, technically a neutral area. The march took Butler through western Georgia and into Florida on the Apalachicola River. From Fort Gadsden, on said river, the march went nearly straight to the heart of the Miccosukee territory near the lake of the same name. There, a brief skirmish with the Indian forces under Kinhage took place and many cows, scalps and other goods were taken. As the enemy fled south, Jackson's forces followed hotly in pursuit. Another skirmish was fought near Suwannee Old Town and many prisoners and cattle were taken. The army then, after the capture of Arbuthnot and Armbrister, two British subjects suspected by Jackson of arming the Indians, held a court martial at St. Marks, Florida. Robert Butler, who had participated in the trial in which both men were found guilty as charged, signed the execution order. The executions caused a major international incident, which was soon smoothed over by tactful diplomacy.

After this exciting and dangerous mission had been completed, Butler was assigned to be the secretary to the commission negotiating with the Chickasaw nation. As secretary, Butler was assigned the duty of delivering the treaty in person to the president, James Monroe. Jackson wrote to the president, "Sir, This will be handed you by Colo. Robert Butler Adjutant Genl of the Southern Division, and who acted as Secretary to the commission charged with holding a treaty with the chikisaw [sic] nation, who I beg leave to introduce to your personal knowledge as a good citizen & valuable officer who has served throughout the whole Southern campaigns in the late war, and accompanied me in the late campaign against the Seminoles, and to whom I beg leave to refer to you for any information on the subject of the latter campaign."[8] Although Calhoun, in his efforts to embarrass Jackson, accused Butler of putting in "exorbitant" rates for his rooms and meals, Butler had the total confidence of Jackson, and later, Monroe.[9] Butler remained in Washington after delivering the papers and kept Jackson fully informed on the actions of his enemies and the passage of the treaty.[10]

Upon returning to Tennessee, Robert Butler began to get his affairs in order. He remained on "duty" as Jackson's Adjutant, a very important post in the turbulent political climate of the day. In early 1819, he began getting Jackson's papers in order and notified the General of certain letter book that had disappeared from the collection. He voluntarily took it upon himself to oversee the organization of

---

[8] *The Papers of Andrew Jackson,* Volume IV, *1816-1820,* 1994, 245.
[9] Davis, 55.
[10] *The Papers of Andrew Jackson,* VI, *1816-1820,* 255-56.

these papers so Jackson could quickly and effectively answer his many critics, especially Calhoun and William Crawford of Georgia, both of whom viewed Jackson as a dangerous political rival for the presidency. During this period, Jackson frequently called on Butler's services to answer his critics and those who may have felt slighted by certain rumors or comments in the press.[11] As an honorary reward for his contributions during the Florida campaign of 1817-18, Butler was given the command of the party overseeing the exchange of flags at St. Augustine, in 1821.

As a reward for his many services, Monroe, on Jackson's urging, appointed Butler the first Surveyor General of Florida in 1824. The tasks before him were of great magnitude. As the government depended upon the sale of public lands for financial support, there was a great urgency in the directives coming from the General Land Office to Butler. After seeing that the confirmed Spanish land grants were laid off and separated from potential public lands, Butler was charged with running out the Prime Meridian, Basis Parallel and laying out the new capitol, Tallahassee. Recruiting some of the finest public surveyors available, Butler had these tasks accomplished in remarkably quick time. However, the very nature of Florida's land, the belief that the rainy season was unhealthy, nay deadly, and the lack of transportation led to a slow start for the public surveys and the sale of usable, cultivatable lands. The large number of "impassable swamps" in Florida made the accomplishment of any survey very difficult. Because the 1820s were, apparently, very wet, many surveyors had to request extensions of time for the completion of their contracts. Additionally, experiments in surveying, such as the "compound meander", were tried but proved ill-advised, taking up additional time and public monies. His task as the first Surveyor General was not an easy one.

After laying out Tallahassee, where he purchased a lot, Robert Butler began to purchase land around the beautiful Lake Jackson, name in honor of the General. The plantation he established was situated on the northern shore of the lake with a beautiful view and fertile land. For some, yet unknown reason, he attempted to sell the plantation in 1833, but, luckily, found no takers and remained on the land until he passed from this world. The advertisement for the land, however, gives an excellent picture of the extensive nature of his holdings:

> Plantation for Sale.—Considerations of private interest in the west, induce me to offer for sale, my tract of hammock land, on which I

---

[11]*Ibid*, 290-91; 400-01.

reside, bordering on Lake Jackson, five miles from Tallahassee. This tract, combining quality, fertility, & beauty of situation, together with extensive improvements, is perhaps not surpassed by any other in the Territory of Florida, and affords at once a residence for a large family, with every convenience suiting taste and comfort. An extensive dwelling-house & kitchen, cotton-gin, and press, corn-mill, sugar-house, stables, etc. etc., of the best materials with a beautiful garden spot, well supplied with choice fruit and shrubbery, and a young orchard in a state of towardness of choice fruits of different kinds, and many springs of delicious water. Nearly three hundred acres in cultivation, and about five hundred under fence, wanting but little repair. Time will be afforded to meet the payments, for a large portion of the purchase money, they being well secured, and an indisputable title will be given the purchaser...Apply to the subscriber on the premises. Robert Butler. Lake Jackson, November 16th 1833.

From the first purchase of portions of this land, in 1825, until just eight years later, all of these improvements were made. The fact that this extensive plantation existed at all at this period is remarkable and speaks to the industry of Butler and his labor force.[12]

This plantation became one of the showplaces of Middle Florida and the center of much of its social life. Ellen Call Long, in her famous *Florida Breezes,* and others, have described the life of the Colonel and his family on his Lake Jackson plantation. Mrs. Long, the daughter of Richard Keith Call, noted in her volume that the roses were the Colonel's pride and joy. His farm included an orange grove (in first bearing), a sugar cane field that stretched to the lake's shore and, of course, cotton fields. She also wrote of the main house as having two halls, one crossing the other, where dances and other social events took place, "detached rooms, called offices," which were where the men retired to discuss business, and other family rooms. The halls were described as "spacious." The main social event of the season was held at Butler's estate and was known as the "feast of roses." The food and drink consumed there were of the finest quality and in ample quantity. Although it was situated five miles from Tallahassee, the social season always included Butler's feast.[13]

Colonel Robert Butler was an active participant in the affairs of the commu-

---

[12]Tallahassee *Floridian*, November 30, 1833.
[13]Long, Ellen Call, *Florida Breezes* (Gainesville: University of Florida Press, 1962. Reprint of 1883 edition), 116-18.

nity. In conjunction with other Masons, he was instrumental in founding the Andrew Jackson Lodge in Tallahassee, on December 19, 1825. He was also a mover behind the founding of the first school in Tallahassee, the sessions being held in the Mason Lodge building. He was also active in the affairs of the Presbyterian Church and was the vice-president of the Agricultural Society of Middle Florida.[14]

It is not surprising that four other of the first masons were U. S. Deputy Surveyors, who served under Butler. Romeo Lewis, later Sheriff of Leon and Jackson Counties, LeRoy May, Davis Floyd and Major Benjamin Clements, who surveyed the Prime Meridian and the Western half of the Basis Parallel, were these surveyors. Included in the first group of surveyors that were hired by Robert Butler, was Henry Washington, considered by the Bureau of Land Management's historian as one of the nations finest and most accurate, the Donelson brothers—Butler's cousins—Charles C. Stone, who ran the eastern half of the Basis Parallel, Thomas White, William McNeill, who ran one of the lines between Georgia and Florida, and R. C. Allen, later territorial Circuit Judge. These men, all whom were leaders in the community in their own right, made up the core group of surveyors in early Florida. Although some "mistakes" did occur, on the whole, these surveyors were remarkably accurate for the first run. Butler, whose integrity was never in doubt, always supported his men when the occasion arose and often encouraged them to exercise their best judgment. From the large volume of correspondence available, it is readily apparent that these men held their leader in high esteem.

Butler was not only the first but also the longest serving Surveyor General in Florida's history. Therefore, he may be judged fairly on the accuracy of the surveys completed under his regime and the lack of controversy about the surveys in general. Overall, professional land surveyors of today give Butler very high marks because most of the early surveys, considering the relatively crude instruments used, are remarkably accurate and can be followed in the field to this day. This holds true, especially, for the work done by Washington, Clements, Benjamin F. Whitner Jr., Arthur M. Randolph and others. Of course, some of the early surveys contained major errors or incorrect directions, however, compared to his immediate successor and other state surveyor generals, Butler still ranks as one of the best.

From the above description of this important man's life, it can be readily seen that his homestead deserves to be placed upon the National Register of Historic

---

[14]Crowther, *The Grand Lodge of Florida Free and Accepted Masons History, 1830-1989*, 12-13.

Places. However, the dwelling house and some of the out buildings were burned in 1886, long after the Colonel had passed to his reward, in 1860. While he lived there, he fathered ten children and provided well for them. Unfortunately, eight of his children preceded him in death and may be buried on the property in the family graveyard. However, the only surviving evidence of the entire family is the gravesite of Robert Butler, whose grave was remarked in 1901 by the Masonic Lodge and another monument was placed there more recently. The Butler gravesite sits just outside of the boundaries of Lake Jackson Mounds Historic Site and on private property just off Crowder Road.[15]

---

[15]Elizabeth Smith, "The Fighting Butlers Come South," *Magnolia Monthly*, Volume 14, No. 2, 1976.

CHAPTER 2

BENJAMIN PUTNAM

On January 11, 1849, Representative John Tanner of Jackson County moved to make a change in the Senate bill entitled, "An Act to Organize the county of Hilaka." What Tanner wanted to do was to change the name of the proposed county to Putnam. When Representative John P. Baldwin of Monroe County sought to strike this, he was defeated in his motion by a vote of twenty-six to two. The rules were then waived and the bill read a third and final time and approved overwhelmingly.[1] Although controversy has been raised over which Putnam the new county was named for, all of the evidence, both contemporary and historical, points to Benjamin A. Putnam as the individual so honored.[2]

What sort of man would deserve such an honor as to have an entire county named for him? During his life, he had many contemporaries who were famous and have since had their reputations enhanced by inclusion in the history books. One needs only mention David Levy Yulee, William D. Moseley, Richard Keith Call, John Milton and inventor John Gorrie to see the point. Yet, with the exception of Senator Yulee, no other "great man" was so honored. To understand why this is so is to comprehend the life and times of Benjamin A. Putnam and the numerous endeavors in which he was involved. It is precisely this that the present study shall attempt to do.

As Putnam County's historian, Brian E. Michaels has told the basic outline of his life in *The River Flows North: A History of Putnam County*, the facts are familiar to many, but need to be repeated, to a limited extent, for a better under-

---

[1] *Journal of the Proceedings of the House of Representatives of the General Assembly of the State of Florida* (Tallahassee: Office of the *Florida Sentinel*, 1848), 197-98.

[2] Brian E. Michaels, *The River Flows North: A History of Putnam County*. Palatka, Florida (Putnam County Archives and Historical Commission, 1986), 74-80. I agree completely with the author's conclusion as to the origin of the name of the county and have followed the same assumption in this paper.

standing. Benjamin A. Putnam was born December 16, 1801, in the lovely port city of Savannah, Georgia. Unfortunately, his father, a former surgeon in the Revolutionary army, died a year later. His name, too, was Benjamin, son of Henry Putnam, who was killed in the Battle of Lexington. He was the grandnephew of another Revolutionary War hero, General Israel Putnam. It would appear that his lineage almost seemed to mark young Benjamin for future military laurels. However, he did, in all likelihood, receive the usual tutoring at the family plantation outside of Savannah, and, at an early age, was shipped off to New York for a year to prepare him for the rigorous curriculum at the famed Phillips Academy in Andover, Massachusetts. The only reason a student attends Phillips is to prepare for admission to Harvard, which is exactly what young Benjamin did.

Young Putnam's stay at the fabled institution on the "Square" lasted two years. This was not uncommon for young gentlemen of the day. A four-year degree seldom opened any more doors than attendance and grooming. The primary purpose was to learn how to research, study and apply lessons for future use. Most of the young planters at Harvard (or any of the schools favored by the elite) attended to acquire these skills and associate with others of the same class. Many of the young planters, though by no means all, went for the practical reasons of learning enough of the law, mathematics, sciences and grammar to be able to communicate well and run a plantation, when the proper time came. Education was seen as a means to very practical ends, aside from the accepted moral training and social graces one acquired while attending.

Upon the completion of his second year, young Putnam returned home to attend to family matters, most notably an ill mother. It is presumed that he moved further south to St. Augustine, Florida in 1823 and studied law in a local office. By the following year, he had become associated with Territorial Judge Joseph Lee Smith, the father of Confederate General Edmund Kirby Smith and the offender of Ralph Waldo Emerson's sensitive ears with his profanity. Judge Smith had the backing of several prominent individuals on the frontier of East Florida. When his term was about to expire in 1832, many of these supporters petitioned President Andrew Jackson to continue his term of office. The list contained the names of numerous well-known leaders and pioneers, including John Lowe, Thomas Ledwith, Edward Wanton, Thomas Flotard and members of the Osteen, Sparkman and Thigpen families.[3] That Judge Smith was also controversial can be

---

[3]Clarence E. Carter, Editor, *The Territorial Papers of the United States:* Volume XXIV. *The Territory of Florida, 1828-1834* (Washington: Government Printing Office, 1959), 597-98.

seen from the petition presented to President Jackson in 1829, defending the jurist from "accusations" made by Joseph Sanchez that were sent to Washington. This same petition contains the names of nearly every recognized leader of East Florida and the name of his protégé, Benjamin A. Putnam.[4]

Putnam's personal attachment to Smith became much stronger in 1830 when he married the judge's sister-in-law, Helen Kirby, in Charleston, South Carolina. She was the daughter of the prominent Litchfield, Connecticut, attorney, Ephraim Kirby. By this time, Putnam had already made a name for himself in St. Augustine. In May 1824, he had been admitted to practice law in the Territory for the spring term of the Superior Court.[5] In a seemingly bold move, he opened his own law office in August of the same year.[6] By 1826, Putnam had become sufficiently respected to be elected Orderly Sergeant for the St. Augustine militia unit known as the Florida Rangers and, in November of that year, was elected to the office of alderman.[7] In 1830, the Putnam's only child, Catherine, always known as "Kate" in the family, was born in St. Augustine. With mother, wife, law practice and reputation established, Benjamin A. Putnam was on the brink of becoming a very important citizen of the Territory.[8]

Putnam's involvement in civic affairs can also be seen in the numerous petitions to Congress and the Territorial Legislative Council he signed. To cite just two examples, on November 8, 1833, his signature is found affixed to the petition to make further repairs to the "King's Road" for twenty miles south of St. Augustine and about the same number of miles north of the Ancient City.[9] And on April 7, 1834, he joined one hundred and forty-five fellow citizens petitioning the federal government for further repairs to Fort Marion for the dual purpose of historic

---

[4]Carter. *Territorial Papers,* Volume XXIV, 293-94.
[5]St. Augustine *Gazette,* May 29, 1824, 2. The announcement also listed one Thomas Correll as being admitted to the practice, however, little is known about Putnam's colleague at this time.
[6]St. Augustine *Gazette,* September 25, 1824. The announcement bears the date of August 21, 1824.
[7]Benjamin Putnam biographical file, St. Augustine Historical Society, St. Augustine, Florida. All of the information through 1837 is taken from the *East Florida Herald.* The author would like to thank Jean Parker Waterbury for providing copies of this file for use in this piece.
[8]Michaels, *The River Flows North,* 80-81. Also see typescript, "Benjamin Alexander Putnam: Biography," by R. M. Burt, on file at the Putnam County Archives. This copy shows the corrections made to Burt's earlier work on its margins. The author would like to express his thanks to Janice Mahaffey, Archivist, for her invaluable assistance in obtaining this and other information contained herein.
[9]Carter, *Territorial Papers,* Volume XXIV, 935-36.

preservation and its utilization as a supply depot.[10] By 1833, he has risen high enough in the social and political community to be appointed Justice of the Peace for St. Johns County, a very important post during the Territorial Period.[11] Service to the community through cooperative petitioning, office holding and charitable work—he was appointed Vestryman in the St. Augustine Trinity Episcopal Church in 1840—brought recognition from his fellow citizens. In 1835, he served his first term in the House of Representatives of the Legislative Council. After attending the session, he became active in soliciting a charter for a railroad between St. Augustine and Picolata from the same legislature. Thomas Douglas and David Levy assisted the young legislator in drawing up the memorial, which was successful in getting the charter during the following session, although it did not prove to be a successful venture at the time.[12]

In this period, the Legislative Council met once a year for about two months, depending on the business to be disposed of. During Putnam's first session, he was appointed to the Judiciary, Banks, State of the Territory and Enrolled Bills standing committees. He served as Chairman of the latter. The issues that concerned him most in this important first session related to his interest in the law and personal property. On the first full day of business, he introduced a petition of Mary J. Fontane asking authorization from the Legislature to sell certain real estate holdings, something normally forbidden to women at the time.[13] He also represented his constituency directly by asking for the incorporation of the Methodist Episcopal Church, and, later, other local churches as well.[14] Putnam also took care to look after some of his legal clients when he presented a petition to exempt Peter Mitchel and other proprietors of the Arredondo Grant in Alachua County from taxation because their specific lands had not been designated by a court of law and, therefore, no one knew exactly what lands they held or their value.[15] He soon observed that many of the acts being proposed by his colleagues were for revision or amendment of acts then in force. Because so much time was diverted to this effort, he felt obliged to move that all such revisions or amend-

---

[10]Carter, *Territorial Papers,* Volume XXIV, 997-99.

[11]Michaels, *The River Flows North,* 80.

[12]Thomas Graham, *The Awakening of St. Augutine: The Anderson Family and the Oldest City: 1821-1924* (St. Augustine: St. Augustine Historical Society, 1978), 33.

[13]*Journal of the Proceedings of the Legislative Council of the Territory of Florida* (Tallahassee, The Floridian, 1835), 13. Hereafter, *House* or *Senate Journal,* year and page number.

[14]*Ibid,* 25.

[15]*Ibid,* 25.

ments be first submitted to the Judiciary Committee for examination and report. This action would free up more time for all members to discuss and devise legislation of more immediate benefit.[16] For Putnam, this meant work on the passage of bills to incorporate the St. Augustine Wharf Company, the partition of real and personal property and a bill to "enable Married Women to convey their real estate of inheritance in this Territory."[17] In this latter bill and his votes in favor of granting divorces, Benjamin Putnam showed a concern for the rights of women that was considerably in advance of his time, although his motives were more based upon legal representation than actual concern for civil liberties.[18]

One of the more important concerns of Benjamin Putnam was that for the benefit of public education. On January 24, 1835, he offered and read the preamble and resolution he had helped to prepare regarding school lands and their disposition. His resolution, which he presented three days later, is one of the more revealing of the day regarding the problem of financing public education in Florida. Also, the sentiments expressed therein show a deep commitment to the ideals of sound government in a democracy:

> Whereas, It is important in all new and growing countries, and especially under a Republican Government, which is based upon the general intelligence and virtue of the community, to provide for the education of the rising generation: And whereas, Florida is almost entirely destitute of the necessary means for that purpose, on account of the sterility of a considerable portion of the sixteenth sections, which have been reserved for the support of schools, and a large portion of the counties having been granted out to private individuals by the British and Spanish Governments before its transfer to the United States, without any reservation of School Lands: And, whereas, also owing to the great extent and conformation of our sea-coast, there are and necessarily must be, a great many fractional townships in this Territory, containing less than sixteen sections-each of which have no School Lands. Resolved, therefore, by the Governor and Legislative Council of the Territory of Florida, That our Delegate in Congress be requested to use his exertions to procure the passage of a law authorizing the selection in such manner as the Governor and the Legislative Council shall direct, other lands in the place of such sixteenth sections as shall prove to be of little or no

---

[16]*Ibid*, 29.
[17]*House Journal, 1835*, 30-47.
[18]*House Journal, 1835*, 30-47.

value, on account of the sterility of their soil, or any other cause. And also the selection and location of a quantity of lands equal to one thirty-sixth of all the lands which may have been granted out as aforesaid by the said British and Spanish Governments, and of the said fractional township.

The young representative would then have the proceeds from the sales of such lands distributed to the counties according to a ratio of the white population in such counties.[19]

His votes on bills pertaining to internal improvements and related topics were also interesting and show him as someone not yet ready to vote for every improvement scheme that came along. For example, he voted against a bill to incorporate the Union Railroad Company and an act to amend an act to incorporate the subscribers of the Union Bank of Florida. Yet, during this same session, he cast his vote in favor of the South Florida Land Company. These votes appear, on the surface, to be contradictory.[20] However, because of a lack of correspondence or diary, the actual causes of his apparent inconsistency must await later analysis. It was in this same session that the act establishing the Southern Life Insurance and Trust Company was passed, yet the roll call vote for this controversial banking house was not recorded and we do not know Putnam's position relative to it. However, in reviewing his first legislative endeavor, Benjamin Putnam had much to be proud of, including the passage of his act for the collection of rents, the incorporation of the Methodist Episcopal Church, the incorporation of the St. Augustine Wharf Company and the act enabling women to convey real estate of inheritance.[21]

The years of 1835 and 1836 were full of adventure, triumph and defeat for Benjamin Putnam. His election to the rank of major in the local militia had to fill him with pride. Yet, he did not return to the Legislative Council in 1836 and was, instead, appointed to be a Notary Public for St. Johns County.[22] The most significant event of his early life, however, was soon to take place at Dunlawton Planta-

---

[19]*Ibid*, 51-52.

[20]*House Journal, 1835*, 105-111.

[21]*Acts of the Governor and Legislative Council of the Territory of Florida* (Tallahassee: William Wilson, 1835). See the listing of Acts in the front of the volume, ii-iv. By reading the listing and reviewing the *Journal of the House*, I matched those Acts passed with those Putnam sponsored from the floor.

[22]*House Journal, 1836*, 114.

tion. Here a single defeat at the hands of the Seminole Indians was to have grave consequences for the remainder of his life.

The disaster at Dunlawton was one of the first notable skirmishes in the Second Seminole War. In actual numbers, the engagement was small and neither side had over one hundred and twenty men. Because, however, it involved a detachment of men from St. Augustine under Major Benjamin Putnam, who were sent to protect the plantations of the area, the short battle received great attention. The most significant fighting took place before the St. Augustine men were forced to retreat because of a lack of ammunition and provisions. Forced to withdraw, the men made their way to the boats that had ferried them to Dunlawton, but found that the tide had ebbed and left their boats stuck in the mud. The men were thus exposed to withering fire from the guns of the Indians as they attempted to free them and took heavy casualties, four dead and thirteen wounded.[23] The loss inflicted on the battlefield later became a political liability to Major Putnam, even though he was later promoted to Colonel and Adjutant General of the Militia.

The reaction of his political enemies was of considerable importance. He was characterized as inept and incompetent in some of the letters written in the aftermath of the engagement. His opponents filed court suits and claims to Congress, which kept the small battle in front of the literate public. His defenders noted that the troops he led were raw militia, relatively undisciplined, who responded as well as could be expected under the circumstances. They also argued that the militia did not have the necessary supplies and ammunition to effectively carry out the task assigned to it. Whatever the case, the young, inexperienced commander, with raw troops, under heavy fire and outnumbered, did the best he could. The only negative fact that cannot be disputed is that he lost track of the time and the cycle of tides, which left the boats and his men greatly exposed to enemy fire. For his enemies, this was enough to condemn him, which they loudly did in nearly every subsequent election.

Putnam did not take part in any other major battles of that war and remained in St. Augustine rebuilding his practice. Somewhat ironically, he represented many

---

[23] John Mahon, *History of the Second Seminole War* (Gainesville: University Presses of Florida, 1985), 137-38. This is the best history of the war and is the standard interpretation of Putnam's battle. I have also consulted other texts, especially the *Autobiography of Thomas Douglas* (New York: Calkins and Stiles, 1856), 118-19. Testimony about the engagement and the damage the war did to the plantations can be found House of Representatives Document No. 92, 27th Congress, 2d Session ("Representatives of Francis Pellicer") and Senate Document No. 129, 25th Congress, 3d Session. (A report from the Committee of Claims)

of those who put forth claims against the government for losses suffered during that tragic conflict. His reputation could not have been so bad as to cause him to lose either clients or elections during the remaining period of the war. His election and service in the 1840 House of Representatives, where he was to be the public voice favoring the division of the Florida Territory into East and West sections, gives sound evidence that his popularity with his St. Augustine constituents was not entirely destroyed.[24]

Throughout the 1830s, 1840s and into the 1850s, Benjamin Putnam was intimately involved in the resolution of the Arredondo claims. These extremely long and drawn out claims and counter claims attracted the best legal talent to be had in the Territory. John Rodman, Joseph Smith, John Drysdale, Thomas Randall and others all appear as legal counsel for one or more of the disputants in these complicated cases. Putnam's major role was in representing Peter Mitchel, the co-owner of the grant along with the Arredondos and Moses Levy, who held the largest amount of "stock" in the Florida Associates, a group of investors in Florida land had formed the association to meet the requirement of getting two hundred families to settle on the Great Arredondo Grant (roughly modern Alachua County) within a three-year time limit set by the conditions of the grant. Others bought into the settlement scheme and soon the project was under way with settlers recruited from New York, Europe and elsewhere. When the Supreme Court of the United States ruled the grant valid under the Adams-Onis Treaty in 1832, the rush was on by the surviving members or heirs thereof to get possession of the land due them by virtue of their investments. The attempts to legally divide the grant among these various shareholders led to seemingly endless litigation, with Putnam representing Mitchel and acting as one of three "appraisers" of the land, when one of the major suits was settled.[25] He also took part in the case of *Brush vs. Prall*, representing the interests of the Mitchel estate and other members of the

---

[24]*East Florida Advocate*, September 21, 1839. This Jacksonville newspaper reported on the meeting of August 29, 1839, where Putnam made an important address favoring division of the Territory into East and West Florida. This was in opposition to the position taken by the Constitutional Convention, which had earlier convened in St. Joseph for the purpose of creating a constitution for a united Florida.

[25]St. Johns County Clerk of the Court Records, under care of the St. Augustine Historical Society. See the case of *Brush v. Prall*. The most important documents concerning the Great Arredondo Grant can be found in boxes 91, 92, 93 and 215. The author, while researching in these documents in conjunction with current litigation, became very impressed with the legal abilities of the St. Augustine attorneys arguing the case. The interrogatories and the responses to them are some of the most informative documents concerning the settlement of the grant.

Brush group of heirs. This case, which ended in the relatively successful division of the Great Arredondo Grant, stands as written testimony to Putnam's great perseverance in the face of very daunting odds.[26]

Putnam was not wholly successful in preserving the interests of the Mitchel estate in the Great Arredondo Grant. In the case of *Benjamin A. Putnam, Executor of Peter Mitchel, vs. John H. Lewis, et ux*, Putnam admitted being taken in by the lack of protest by Lewis to the estate claims of Peter Mitchel's heirs to land within the grant. Putnam, however, appears to have been too trusting of Lewis. As the court report states: "The petition alleged that petitioner had been lulled into security by the letters and declarations of John H. Lewis, that he had no design to assail the claim of Peter Mitchel, his testator, until about the time the cause came on for hearing, when it was too late to make any effectual resistance: That being taken by surprise when the whole claim of Peter Mitchel in the said grant was most unjustly and unexpectedly assailed by said Lewis and wife, he was force to submit to a compromise with the said Lewis and wife; and that he, under the circumstances, had consented to the said decree." The case was somewhat of an embarrassment to Putnam in that his lack of diligence costs his clients a large number of acres in the grant. In attempting to recoup his loss, Putnam appealed the decision until it reached the Supreme Court in the January Term of 1847. In opposition was the highly respected and able Thomas Randall. After many pages of legal arguments and citations to authorities, the Supreme Court decided against Putnam and upheld the lower court's ruling, with costs.[27]

Another case argued by Putnam in front of the Florida Supreme Court was styled, *John B. McHardy and the Creditors of Robert McHardy, Dec'd., Appellants vs. the Surviving Executor of Robert McHardy, Et. Al., Appellees*. This case, which began in the Circuit Court for East Florida, St. Johns County, pitted Putnam against George R. Fairbanks (a former clerk in the office of the Superior Court) and the young George W. Call. Greatly simplified, the case revolved around the issue of who should be paid first from money obtained from Congress due to losses to McHardy's estate during the "invasion of the Province of East Florida"

---

[26] Records in the holdings of the Clerk of the Court for St. Johns County, and archived by the St. Augustine Historical Society, St. Augustine, Florida. See Boxes 91 and 215 for most of the documents regarding this long, complex and vexatious litigation.

[27] *Florida Reports,* Volume 1 (Tallahassee: Samuel S. Sibley, 1847), 455-76. The case was actually decided on the technicality that you could not appeal a decree of a lower court that was interlocutory. It recommended a bill of review be brought in the lower court, however, I have not found where this was done and can offer no explanation for this inaction.

in 1811-13. The court ruled in favor of the clients of Fairbanks and Call because they represented the legal creditors of the estate where money was actually set aside to pay all legal and just claims, regardless of the source of the money. Putnam had the unenviable job of attempting to argue that the money was paid for damages and belonged to the rightful heir(s) and that the debts of the estate must first come from any other gains or profits made by the sale or use of the property of the estate. The court, while ruling against Putnam and his clients, nevertheless complimented him on his professional ethics in not insisting upon an argument that would have greatly delayed the court's decision making process and cost the estate considerable money in attempting to justify its shaky legal position.[28]

In one of the more interesting cases in which Benjamin Putnam was involved, Phoebe, a slave woman belonging to the Anderson family was accused of attempting to poison a Mr. Landon, to whom she had been leased. Mrs. Anderson appears to have had a strong personal attachment for the woman and hired Putnam to defend her in the winter term of the circuit court in 1842. Putnam was required to go to Jacksonville to accomplish the task of acquitting the beloved slave. However, as historian Thomas Graham noted, "Mrs. Anderson's lawyer, Major Putnam was not at all sure that Phoebe was innocent, despite the verdict of the court. 'She made a very narrow escape,' reported Putnam, explaining that the evidence against her was substantial." Phoebe, much beloved and chastened, was later sold in Charleston in 1851.[29]

Returning to the legislative career of Major Putnam, his record in the 1840 session gives clear indication where his future political allegiance lay. During this session, he served on the Committee on Internal Improvements and the Judiciary Committee. The crucial issue of banks took precedence in this legislature and the debates must have been more than tepid. On January 23, 1840, Putnam moved the resolutions put forth in the House be placed upon the calendar for the following day, at which time the Senate resolutions concerning investigations into Florida Banking practices came to the floor. In both cases, Putnam voted in the negative, in opposition to the investigations and resolutions. On February 25, he signed a petition to the Speaker that questioned the legality of the legislatures investigation into the affairs of the Union Bank of Florida. He also voted against an act to amend the charter of the Bank of Apalachicola. Loosely construed, the position of

---

[28]*Florida Reports,* Volume VII, No. I (Tallahassee: James S. Jones, Printer, 1857), 301-18. See, especially, the court's comments on 308-09.

[29]Graham, *The Awakening of St. Augustine,* 79.

Putnam in relation to the banking houses was favorable to their maintenance. This, of course, was exactly the position of the nascent Whig Party of Florida, which was then being formed.

Another position taken by many of the soon to be called Whigs, was in opposition to free or reduced costs of public lands. In the only major resolution to be introduced in this legislature related to preemptions, Putnam was found in opposition. Additionally, his vote regarding the bill to suspend the revenue laws for 1840 put him in opposition to the majority that favored such a move. Both of these votes showed the "conservative" side of Putnam and placed him firmly in the ranks of the future Whig Party. As one of the leading spokesmen in the legislature against these measures, his statewide reputation grew.[30]

The position taken by Putnam in regards to the division of the Territory is also reflected in this session. Early in the session, he had voted with the minority in an attempt to table the motion to transmit certain documents related to the state of the Territory to the U. S. Senate. He supported the minority report of his colleague K. B. Gibbs in opposition to statehood, which opposed the majority report, which stated that a "large majority of the people are opposed to any division of the said Territory." And Representative Putnam also voted with the minority to attempt to get the minority report sent to Congressional Delegate Charles Downing.[31] Along with many of his East Florida constituents and his colleagues in the legislature, Putnam was in favor of the division of the Territory and opposed to statehood. When elected to the last session of the Territorial Council as a Senator from St. Johns County, he would lead the last ditch fight against statehood.

As a Senator from the Eastern District, Putnam was elected to the 1845 session and was assigned to the Judiciary, Schools and Colleges, and Propositions and Grievances Committees. He was also selected to serve on the Joint Select Committee to report on those portions of the Governor's message pertaining to Faith Bonds and Guarantees. When he asked to be excused from service on this Select Committee, since he was the token minority voice, he was denied. The majority was from the ranks of the rapidly developing Democratic Party and Putnam and

---

[30] *Journal of the Proceedings of the Legislative Council of the Territory of Florida* (Tallahassee: J. B. Webb, 1840). Putnam's votes on the issues discussed in paragraphs concerning this session can be found on pages 49, 86, 100-01, 144, 150, 156-57 and 175. It is interesting to note that the majority of the House voted not to amend the charter of the Southern Life Insurance and Trust Company in which many of Putnam's closest St. Augustine associates, including Judge Smith and Peter S. Smith, were officers or major stock holders. (See pages 46 and 156-57.)

[31] *House Journal, 1840*, 23, 77-78, and 125.

his colleagues from East Florida knew they were fighting a losing battle. However, this did not prevent them from attempting to derail the train to statehood at every opportunity. Amendments were offered and defeated, postponements were voted upon and lost and resolutions were read, but not put on the record. The most telling document related to the division issue was written by Putnam with assistance from I. D. Hart and Representative H. H. Phillips of Duval County and which contained most concise and clear arguments against statehood at the time and put forth the position of most of East Florida's voting population.[32]

This report immediately launched into the reasons why there should be a legal division of the Territory. Putnam argued that Spain had initially set up the Territories into two separate political entities, East and West Florida, with two distinct governments. The British, when they acquired Florida in 1763, kept this arrangement intact, as did the Spanish when they reacquired the Floridas in 1783-84. When General Andrew Jackson took over the Territory of Florida on behalf of the United States in 1821, he too, recognized the separate regions of the Territory by appointing two separate officers to govern each district. After putting forth the historical basis, he continued by noting that the people of East Florida had continually supported division of the Territory, even as late as 1844, in a resolution to Congress. There was nothing in the previous legislation of the Federal Government that could be construed, he pleaded, so as to enjoin the two Territories into one state. After stating the legalistic case, the senator then got to the political meat of his subject. The question of balancing the federal government between slave and non-slave states was uppermost on many people's minds and Putnam flatly states, "Whereas, Present indications admonish us, in the most significant manner, of the necessity of preserving a just balance of power or influence between the slaveholding and non-slaveholding States, and make it most manifest that the true interests of the South generally, as well as of Florida, require that the Floridas should come into the Union as two States whenever they are admitted." To assist the South in this struggle for the balance of power, therefore, was one of the most important reasons for division. Specific to East Florida, Putnam clearly indicated that the territory was not in a condition to join in statehood because of the tremendous losses sustained as a result of the Indian war. The higher tax burden needed to repair the damage could not be expected to come from Florida

---

[32]*Journal of the Senate, 1845*, pages 53, 69, 71, 74-75. The Minority Report can be found on pages 53-61.

itself, it needed to come from the federal government. Statehood, if passed, would place this burden squarely upon the shoulders of those least able to afford it. Finally, he argued that the decision for statehood belonged to the people of each Territory—implying that votes from each Territory should not be combined. He noted the familiar argument of the day that the constitution agreed to at St. Joseph in 1838-39 was not ratified by the people of the Territory and that there were too many unanswered questions regarding the tabulation of the votes. As he stated the case, "an instrument of so much importance as this, organizing a permanent government for the people of Florida, should not be forced upon them, when the contest for its adoption has been so warm, the disapprobation of the instrument itself so strongly evinced, and the issue so doubtful."[33] By drafting and presenting this Minority Report, Putnam placed himself in the leading position opposing the adoption of statehood.

The battle over statehood had been long and hard and left many with bitter feelings, which carried over into the elections for the first statewide offices. The old Dunlawton disaster was completely replayed. One newspaper, the *Florida Herald and Southern Democrat* of St. Augustine simply put "PUTNAM AND DUNLAWTON—Boys, remember them when you vote."[34] The candidates were all men well known to Floridians at the time. Richard Keith Call and Benjamin Putnam made up the ticket for Governor and Congressional Representative for the Whig Party and were opposed by William D. Moseley and David Levy Yulee for the Democrats. The election, itself, has been viewed as a contest between only Call and Yulee; however, this runs in the face of the facts and the reporting of the day. Neither Moseley nor Putnam were "unknown" outside of their districts, both men having been leaders in their respective groups and were often quoted in the press throughout the Territory. The highly partisan press of the era had a field day reporting on various debates among the candidates. On one occasion, at a meeting in Newnansville, the *Florida Herald and Southern Democrat* reported that, "Maj. Putnam was completely demolished, used up, by Mr. Levy." The debate, it may be noted, with all of its speakers and hoopla, lasted ten full hours.[35]

The election marked a low-point for the Whig party in Florida. It was poorly

---

[33]*Senate Journal, 1845*, 53-61. For a reading of many of the documents from which Putnam drew his basic philosophy, see Dorothy Dodd, *Florida Becomes a State* (Tallahassee: Florida Centennial Commission, 1945).

[34]*The Florida Herald and Southern Democrat*, 20 May 1845, 2.

[35]*The Florida Herald and Southern Democrat*, 20 May 1845, 2.

organized and lacked newspaper support in the Territory. It should be noted that the Whigs did not meet in any statewide convention but satisfied themselves with District meetings. The nomination of Call and Putnam appears to have been very haphazard and was thrown together with geographical balance more in mind than party compatibility. Indeed, historian Arthur W. Thompson characterized the nomination process as "informal". Without a true party organization and little newspaper support, the fate of Call and Putnam was predictable. The Democrats scored a landslide victory and captured most of the important offices throughout the State. Call and Putnam carried majorities in only four counties, all west of the Apalachicola River.[36] Putnam could only return to his home in St. Augustine, where, in partial recognition for his leadership, he was elected mayor in the November 1845 election.

The new state having been established, Putnam and his colleagues went about organizing the Whig Party of Florida. Little is known about the organizing process; however, it was to prove very successful in 1848. In this year, the Whigs took most of the major offices from the Democrats and returned Benjamin A. Putnam to the State House of Representatives, where he was immediately elected Speaker, with no opposition votes recorded. Politically speaking, this was Putnam's greatest triumph. During this session, he continued to vote conservatively and was frequently in the majority. He did, however, vote against some revisions to the new constitution, including the bill to limit the number of years a judge could sit. In this case, he strongly favored retaining judges for good behavior, while the bill proposed to limit the term to eight years. He was outvoted by a thirty-three-vote margin.[37]

In the following year, Benjamin A. Putnam was rewarded with an appointment to the job of Surveyor General of Florida and set to work at his office on the corner of Artillery Lane and Hospital Street. This was an extremely important position and required long hours and attention to detail. The job was very large in scope because nearly half of the State of Florida was still unsurveyed at the time. There had been many questions raised on the national level about the accuracy of

---

[36]For the best discussions of this election, see Arthur W. Thompson, *Jacksonian Democracy on the Florida Frontier* (Gainesville: University of Florida Monographs, No. 9, Winter 1961) and Herbert J. Doherty, Jr., *The Whigs of Florida, 1845-1854* (Gainesville: University of Florida Monographs No. 1, Winter 1959). Both men viewed the election as a Call versus Levy race. This author does not agree that this is precisely the case.

[37]*House Journal, 1848*, 42. Also, see the entire *Journal* for all of the other votes. Considerations of time and space preclude a lengthy discussion of his role or voting record.

the work of Florida's surveyors and strong scrutiny of their work was a necessity. This job became even more important in the following year when Congress passed the Swamp and Overflowed Land Act. This act was to prove to be the most important legislation ever passed for the benefit of Florida. The act simply stated that any surveyable section of land that was fifty percent or more of a swamp and overflowed nature, but was capable of being artificially drained and made cultivatable could be patented to the State by the federal government. This would require a method of selecting such tracts, hiring competent surveyors to measure and decide on the proper selections and petitioning the General Land Office for the patents. This was a very complex and controversial matter and had to be handled with the utmost care and delicacy. As the first Surveyor General to oversee the execution of this act, there were many details to be worked out and directions to be issued. Putnam, despite the attendant criticism, performed this task well. He also had to continue the controversial surveys of confirmed Spanish land grants, which entailed no end of squabbling. Finally, while performing his regular duties, he had to oversee the surveys in the area of the line drawn by General William J. Worth designating Indian Territory in southern Florida. With pressure from the U. S. Army and citizens eager to have the Indians removed—and attempting to work with the new Board of Internal Improvements—Putnam's days as Surveyor General of Florida were filled with hard work and great tension.

Following his term as Surveyor General, which ended in 1854, he served as Circuit Judge and was, again, elected Mayor of St. Augustine. Putnam also returned to private practice at this time, having his offices on Picolata Street.[38] He was originally appointed to complete the term of Judge Forward, and was subsequently reelected to the post in 1860. He served in this capacity until 1868. As the War Between the States approached, the Judge was looked to for leadership and his every move followed by the St. Augustine *Examiner*. Indeed, the local press was quite laudatory, exclaiming, "Judge Putnam, a resident of this place, discharges his duties with much ability, and with great satisfaction to the whole Circuit. Our citizens may well congratulate themselves in having so much legal attainment, impartiality and integrity on the bench. We certainly have no occasion to regret our system of elective judiciary."[39] By mid-May his travels in the Circuit were completed and the family, for reasons of health and relaxation, looked

---

[38] St. Augustine *Ancient City*, February 23, 1856. See advertisement on the bottom of page 3. The practice of having a judge also taking on private cases was not all that uncommon in this era.

[39] St. Augustine *Examiner*, March 31, 1860, 2.

forward to their planned trip to Niagara and Upper Canada.[40] By the beginning of September, the Judge and his family had returned from their sojourn to New York and Canada, rested and refreshed.[41] However, with the storm clouds upon the land, Putnam saw that others were looking to him for his opinion, which he gave to the St. Augustine *Examiner* on December 13, 1860. The paper reported the following, "It is a source of much pleasure to us to be enabled to State that, this Gentleman is with the South in the present momentous times. He gives it as his belief that the free and slave States cannot possibly continue together, owing to incompatibility of sentiment, and that our only hope of preserving our rights and liberties is in an immediate withdrawal from the present Union, and the formation of a Southern Confederacy." This closing of the ranks was important for all concerned. But Putnam went further than most in his support of the cause. On February 14, 1861, Judge Putnam donated four hundred dollars to the State's coffers. As G. W. Means informed Governor Madison S. Perry, "Judge Putnam desires that the donation be expended by you in such way as may in your judgment be most conducive to the interests of his adopted, though much loved 'Land of Flowers'."[42]

The cool day in January, when the results of the Secession Convention became known, bode ill for the fortunes of the new Confederacy. Colonel Putnam, now in his capacity as local commander of the St. Augustine Blues, read a speech officially announcing the decision for independence. The War would not treat the commander well. During the conflict, he was forced to seek shelter away from St. Augustine in Madison. In 1863, his beloved wife, who stayed behind with her older sister, the "notorious Mrs. Francis Smith," was sent, with other Southern sympathizers, to Hilton Head, South Carolina. The Putnams, Gibbs and other families had already been reduced to "living in windowless shack, subsisting on public charity," to use Thomas Graham's description. At the end of the War, Putnam's property was confiscated and sold to freedmen, forcing the judge to write advertisements in the local newspapers warning against people purchasing land under these claims. It was a very bitter experience for the proud Judge and his

---

[40]St. Augustine *Examiner*, July 7, 1860. This edition reported the family as leaving on the tour. The May 19 edition of the same paper noted the end of the session and Putnam's return to St. Augustine.

[41]St. Augustine *Examiner*, September 1, 1860.

[42]Florida Department of State, Division of Library and Information Services, State Archives of Florida, Record Group 101, Series 577, Carton 1, Folder 8. G. W. Means to M. S. Perry. February 14, 1861.

family.[43]

As a judge, Putnam was noted for his fairness on the Bench to all who came before him; however, after the bitterness of the War Between the States, which saw him move to Madison and have his relatives removed from St. Augustine by federal troops, his judgment may have been prejudiced. Most notably, in the murder trial of James Denton of Micanopy, he presided over a case that does not speak well for the justice of the day and adds to the negative stories of Reconstruction. In this famous case, Denton had openly murdered a freedman named Alec Johnson on the pretext that Johnson had acted insolent toward him. In Putnam's court, Denton was found guilty of manslaughter and sentenced to only pay court costs of $225 and serve one minute in jail.[44] This incident in the Reconstruction of Florida does not speak favorably to the man who had done so much for the benefit of his constituents and state.

After the war, with his land in St. Augustine confiscated and sold to freedmen, Benjamin Putnam moved his residence to Palatka in the county named for him. Here he not only continued as a judge, but experimented in horticulture, especially the loquat, or Japan plum. Like another Putnam County resident, F. L. Dancy, his work in experimenting with new plants helped to establish Putnam County as a leader in the growth of citrus and related crops. It was reported that these experimental seeds lasted and produced for over fifty years after the judge's death on January 25, 1869.

What can be said about a man who served in so many capacities throughout his life? As a lawyer, legislator and judge, he helped to mold the judicial system of his state and territory. As a soldier, he led troops into battle and bravely withstood the onslaught of the enemy. As a politician, he fought his battles with the weapons of the day, admitted defeat when that was necessary and moved on to address other issues. As Surveyor General in a pivotal period, he helped to shape the landscape and mold the future of Florida. The very essence of the man was a boldness of action, an admirable honesty and a willingness to take risks he thought would benefit his constituents and his family. When taken as a whole, the life of Benjamin Alexander Putnam mirrored the age he helped to shape. By almost any measure of man, he was a statesman in the land.

---

[43]Graham, *The Awakening of St. Augustine*, 84-133. This section of Graham's fine work gives many details of life in St. Augustine under Union rule. The portions on Putnam, cited in summary here, are well documented and written in a very readable style.

[44]Jerrell H. Shofner, *Nor Is It Over Yet* (Gainesville, University Presses of Florida, 1974), 88.

CHAPTER 3

A FINDER OF MANY PATHS: JOHN WESTCOTT AND THE INTERNAL DEVELOPMENT OF FLORIDA

Inventor, Surveyor General of Florida, railroad president, canal builder, educator, doctor, soldier, politician, etc. the list of accomplishments of the remarkable John Westcott continues to grow as one investigates his life. He was, as with most men of originality, frequently embroiled in arguments, political turmoil and the search for answers. At the time of his death, he was working on a book concerning the origins, care and cures for yellow fever.[1] His mind seemed never to rest and intrigued all who knew him. He was also a practical man given to attempting the possible, even when the odds seemed forbidding. Because so many of his interests became major avenues of travel and commerce still in use today, John Westcott deserves a closer look as one of Florida's pathfinders.

John Westcott, unlike his famous brother James D. who was born in Virginia, first saw the light of day in New Jersey on June 16, 1807. This date has eluded some who have written of his life, such as Rowland Rerick, who stated that he was eighty-four at the time of his death on December 31, 1888.[2] Little is known of his early education except that it was good enough for him to obtain an appointment to West Point, at the age of sixteen. However, he did not last long at the Military Academy, being admitted as a cadet on July 1, 1823, and resigning by November 15th of the same year.[3] Though it has not been stated in any reference, it has been

---

[1] Quoted in Creville Bathe, *The St. Johns Railroad: 1858 to 1895* (St. Augustine: Allen, Lane & Scott, Philadelphia, Centennial edition, 1958), 56.

[2] Rowland H. Rerick, *Memoirs of Florida*, Volume II, (Atlanta: Southern Historical Association, 1902), 158. Rerick also has his death listed as January 1889, which may be excused in that it was reported in newspapers of that year up to two weeks after the actual event.

[3] Letter of March 1, 1950. Colonel R. S. Nourse, Adjutant General's Office, U. S. Military Academy, to Alfred Hanna, Gilbert Youngberg Papers, Box 4, Rollins College Archives and Special Collections, Winter Park, Florida.

presumed that young Westcott soon entered into the study of medicine, probably in Philadelphia, but this has not been confirmed. That he became a doctor, however, is well known and he left the northeast to journey to Florida, settling in Madison, probably in 1838 or 1839.[4] If this dating is correct, Westcott arrived in Madison at the age of thirty-one or thirty-two and during the conflict known as the Second Seminole War. The time between his leaving the Military Academy and his arrival in Florida during this crucial war remains a mystery because of a lack of documentation.[5]

Westcott soon made a name for himself as a medical doctor and served in the Seminole War with Colonel William Bailey's First Regiment of Florida Mounted Volunteers. He was first mustered into the service on April 20, 1840, as an assistant surgeon, the normal starting rank for a trained physician in the military. He was promoted to the rank of surgeon on August 7, 1840, and remained at this rank for the remainder of his unit's service, which ended in early 1842, the usual enlistment period for Bailey's regiment being three months. Some of his colleagues during his service included Lt. Colonel F. L. Dancy, who was to succeed him as Surveyor General of Florida, Elias E. Blackburn, John Osteen and John H. Gee, of the successful Gee family of Gadsden County, and other prominent pioneers of early Florida. He frequently operated without an assistant surgeon until the appointment of John H. Gee in December of 1840.[6] After the war, he remained in Madison somewhat raised in stature and always referred to as "Doctor Westcott" by local residents.

From March 15, 1844, until August 21, 1845, John Westcott served as postmaster at Madison, an important local position.[7] As post-master, he received the

---

[4]*Applications for Employment*, Volume 2, 1845-56, 57. In a letter of September 10, 1847, S. J. Perry wrote that he had "intimately known" Westcott for eight years, thus giving some evidence of Westcott's arrival prior to 1840. This volume is located in the Land Records and Title Section, Division of State Lands, Department of Environmental Protection, Tallahassee, Florida. Further citations will reference only the *Applications for Employment*.

[5]In a letter of November 4, 1949, Margaret Borland, a descendant of the Westcotts, wrote to A. J. Hanna that only a few medical pamphlets, which her aunt gave away, the family Bible and one piece of correspondence were all that she could trace in her family's records and possessions. Youngberg Papers, Box 4.

[6]"Florida Militia Muster Rolls: Seminole Indian Wars," Volume I, Special Archives Publication Number 67, Florida Department of Military Affairs, State Arsenal, St. Francis Barracks, St. Augustine, Florida. 67-73. ed. Robert Hawk, no date of publication. Copies provided by the Executive Assistant for Government and Community Relations, Ramelle Petroglou, without whose assistance some of this valuable military information could not have been located.

[7]Elizabeth H. Sims, *A History of Madison County, Florida,* (Madison: Madison County Historical Society, 1986), 209.

circular from Thomas Baltzell, President of the Board of Trustees for Seminary Lands, requesting information regarding the state of education in Madison County. Westcott took this opportunity to put forth a plan for the use of the fund and a system of education that was very modern in its methods. In this plan, the Madison post-master put forth the idea of a large common school fund generated from the rents of Seminary lands. The local section sixteen lands, would also be rented out and under local control through a commissioner of school lands located in each county and appointed by the governor and with Senate confirmation. He also called for a local property tax to aid in the support of education which would be voted on by local residents (property holders). Another innovation was the call for the creation of a Secretary of School Land and funds, to reside in Tallahassee, who would oversee most of the bonding and interest payments due to the various counties' school funds. Most importantly, as an adjunct to proper education, Westcott called for "uniform libraries" accessible to all classes in the community. These libraries, he believed, would be "important auxiliaries of public instruction and moral reformation." In a telling phrase, he summed up his notions concerning public education: "Education is to the Republican body politic, what vital air is to the natural body; necessary to its vary existence, without which it would sicken, droop, and die."[8] The calls for a local property tax, support for public libraries, a "Secretary of Education" and rentals of lands for purposes of raising funds for education were far in advance of most other plans put forth for education in the South or elsewhere. There was nothing elitist in his plan and it indicates a Jacksonian frame of reference that was to last throughout his life.

Westcott did not let his plan sit idle while policy makers elsewhere made up their collective minds. In January of 1845, Westcott and other Madison leaders formed Masonic Lodge No. 11. The Masons, with a long tradition of republicanism and public spirit, pushed forth in Madison and soon formed the St. Johns Seminary of Learning, the "best known and most important of these antebellum schools." Located on the eight acres of land acquired by the Masonic Lodge at the corner of Base and Duval Streets, the school was convenient for the students of Madison and the surrounding countryside.[9] This pioneering school offered, according to historian Elizabeth Sims, the equivalent to a high school degree.

---

[8]Nita Katharine Pyburn, "John Westcott's Plan for Public Education in Florida, 1844," *Florida Historical Quarterly*, 27 (January 1949), 300-07. An original copy of this report is also found in the Land Records and Title Section, Division of State Lands, Department of Environmental Protection, Tallahassee, Florida. (File - Rectangular file box, "School Land Selections")

[9]Sims, 38-39. Page 50 notes Westcott's role as a charter officer in the Masonic Lodge.

The respect John Westcott had developed soon led him, like his older brother, into politics and election to the House of Representatives of the State of Florida. Appearing in its first session, in November of 1846, Westcott had an opportunity to lead the new state in many directions. He was made chairman of the House Rules Committee and the Committee on Schools and Colleges, and was a member of the Committee on Amendments and Revisions to the Constitution and the Committee on Printing. Most significantly, he was a member of the Judiciary Committee, charged with the setting up of the new judicial system for the state. In this capacity, Westcott pushed for the concept of elected judges for probate and co-sponsored the bill for "certified copies of records evidence" which would prevent fraudulent documents being admitted as evidence in trials, a concept in use to this very day.[10] From the *Journal of the Proceedings for 1846*, it appears that the Judiciary Committee assignment took up most of Westcott's attention in this session.

However, because education was also an important subject to Westcott, one of his more important assignments was the Committee on Schools and Colleges. In this capacity, as in the plan noted above, Westcott led the way in delivering what appears to be an impassioned address concerning the need to fund public education. All were agreed and he began reading of the committee report, on the need to address the topic and its vital importance. However, the topic had been discussed to such lengths that too many were growing up in ignorance of their rights because of a lack of educational facilities. "A people cannot properly exercise their rights," he declared, "and discharge their duties, without understanding them. In a republican government it is the duty of every citizen to have a knowledge of his rights and duties, and of the means or laws securing the one and enforcing the other. And from this, we derive the obligation of the whole, to furnish the means of study, or in other words provide schools, where rights and duties shall be taught." To fund these needed schools, the committee felt that the immediate sale of the section sixteens would be the best method and from the proceeds to establish a common school fund under a common superintendence. This report, controversial by its nature, was adopted.[11]   A separate Secretary of Education was not set up by the final enactment of the Legislature and control of the lands,

---

[10]For Westcott's support of elected probate judges see, *Journal of the Proceedings of the House of Representatives of the Second General Assembly of the State of Florida at its First Session* (Tallahassee: Southern Journal Office, 1846), 39. For the committee's recommendation regarding the certified documents and voting thereon see pages 44 and 50 in said *Journal of the Proceedings*.

[11]*Ibid*, 62-64.

especially the sixteenth sections, went, instead, to the Register of Public Lands, under the supervision of the Governor. The funds, beyond expenses, were to be invested in interest bearing bonds or like securities. These provisions angered Westcott and fellow representative Charles Russell of Benton County, with Westcott voting against the bill when it reached the floor.[12] The final law, Chapter 93, Laws of 1846, also did not include a provision proposed by Westcott concerning the method of payment over time and required in its place the payment of one quarter down with equal payments, plus interest, following in the next three years.[13] This principle, the payment over an extended period of time, was dear to Westcott and he often proposed it in some form each time the public was required to pay a large amount for property. In this belief, he was in keeping with many progressive thinkers of the day, including Thomas Hart Benton, who championed this principle in Congress.

Rowland Rerick noted that Westcott, while in the legislature, "was one of the projectors of the internal improvement system of Florida."[14] This is not a strictly accurate statement if one includes the promotion of ferry crossings and other such undertakings as incorporated into the realm of internal improvements. Indeed, Westcott opposed such uses of government power to grant monopolies to private individuals. In keeping with his basic Jacksonian principles, John Westcott constantly voted against ferry monopolies or toll roads. His votes in favor of reducing the burdens of financing land purchases show that he favored the "little man" in his politics, as later he would openly oppose the vested interests of St. Augustine and fall heavily into political disfavor with the large land owners who were frequently united with the banking houses. Westcott's support of the concept of internal improvements came from his belief that it should be done by individuals or public corporations, not government subsidies to particular interest groups. Indeed, he voted against the bill for the sale and use of lands set aside for the benefit of internal improvements that became Chapter 94, *Laws of 1846*.[15]

---

[12]*Ibid*, 145-49.

[13]*Acts and Resolutions of the General Assembly of the State of Florida. Second Session, 1846* (Tallahassee: Office of the *Floridian*, 1846), 47-49.

[14]Rerick, Volume II, 158.

[15]*Journal of the Proceedings,* 172-73. A close examination of the Journal reveals his constant defense of the time payment method and attempts to get it approved as an amendment to each bill involving land purchases. His approach is consistent with the national movement for a graduated price for public lands and, eventually, free homestead lands for settlers, first reflected in American public life by the Armed Occupation Act of 1842, sometimes called the "Florida Donation Act" by some historians.

Whether his policies or his constant negative voting in the legislature discouraged him or not is unknown, however what is clear is that John Westcott did not serve in the next legislative session from Madison County. In August of 1847, John Westcott embarked on a career in public lands surveying as a means of supplementing the meager income he made from being one of two doctors in a small town. In his quest for a position as U. S. Deputy Surveyor, he enlisted the aid of S. J. Perry, another surveyor and political leader in Madison County. Robert Butler, then finishing a second tour of duty as Surveyor General of Florida, hired Westcott on August 26, 1847, and brought a sigh of relief from the ex-legislator.[16] By November 26, 1847, Dr. John Westcott was now U. S. Deputy Surveyor Westcott and operating in the not so friendly confines of Township 23 South, Range 24 East, near the headwaters of the Little Withlacoochee River, known today as the Green Swamp. There, with his men and supplies, he gratefully experienced bad weather, swamp and sickness that were the common experiences of surveyors then, and now.[17]

Almost every U. S. Deputy Surveyor who served in Florida, and other territories, was a pathfinder. Few, if any, persons traveled the lines blazed by the surveyors until the time of settlement, which, in some parts of the State, did not come until the verge of the Twentieth Century. In his first surveys in the Green Swamp area of central Florida, Westcott was one of the first human beings, including Indians, who saw the pristine nature of this fabulous resource, which is the source of four major rivers (The Withlacoochee, Hillsborough, Peace and Kissimmee). The territory was one of vast swamps, some of which were unsurveyable at the time, and countless lakes, bayous, baygalls and intermittent streams. Westcott's description of the land is worth noting:

> The work has been tedious on acct of numerous and Large swamps. The head of "Devore" & other Creeks tributaries of the Withlacoochee   Through T 25. R. 23 T 25 R. 24 & T. 24 R. 24 the Main Stream of the Withlacoochee passes, with an immense swamp from 1/2 to 3 miles wide. The South Branch of the Withlacoochee & Hillsborough river rise in Large Swamps...I have carried my "bed & board" on my back & Stomped on the Line upwards of 60 nights, that I might commence my work in day light & work until sunset. I do not mention

---

[16]*Applications for Employment*, 675.

[17]*Letters and Reports to Surveyor General, 1825-1847*, Volume 1. Letter of December 31, 1847, 875. Land Records and Title Section, Division of State Lands, Department of Environmental Protection, Tallahassee, Florida. Hereafter, *Letters and Reports*.

this because there is any novelty in it, but to show the proper perseverance & intensity, has been used; the matter of the country is such that it is impossible to Pack to advantage.[18]

In such primitive conditions the legislator, the community leader, this man of medicine and indomitable will forged forth to create a pattern of lines useable by settlers yet to come.

The difficulty of the terrain was not the only difficulty he faced as a Deputy Surveyor. In addition to the usual problems of Indian scares, bad weather, illness, lack of supplies and poor communications, he also faced the irritations of the bureaucracy in St. Augustine and Washington. One of the many controversies he encountered was a disagreement with fellow surveyor A. H. Jones, a former engineer in many canal and railroad ventures in Ohio and Pennsylvania. Jones, it appears, was under the impression that Westcott wanted to take nearly half of a proposed contract between Jones and the new Surveyor General of Florida, Benjamin Putnam. In two lengthy letters to Putnam, Westcott tried to persuade him that there was enough territory for at least two contracts and that the survey of those in the vicinity of the Fenholloway and Steinhatchee Rivers headwaters was of vital concern since the Indians, in the previous war, had used this area to the great harm of nearby settlers and soldiers. Putnam, however, had given his word to Jones and offered Westcott a choice of territories to survey, thus avoiding a difficult situation. Westcott chose the area near Peace River as his next contract, the other offer being "scrap work" in western Florida where he had no information as to the lay of the topography or distances between scraps.[19] Although Westcott had an additional contract for the 1849-50 surveying season, he also had hopes of more involvement in politics for he noted to Putnam, "My friends here Say my Election is Certain, knowing the uncertainty of Such matters, I am not So yet Sanguine, yet feel that the probabilities are in my favour."[20] His election, as he feared, did not result in a seat in the legislature and on December 3, 1850, he wrote to Putnam that, "I am now So Situated that I could commence operations immediately."[21] One does not ask for a surveying contract at a time when the

---

[18]*Letters and Reports*, Volume 2, 273. Letter of March 30, 1848.
[19]*Applications for Employment*, Volume 2. 683-88. Letters of October 13, and November 1, 1849; and *Letters and Reports*, Volume 2, 291-92.
[20] *Applications for Employment*, Volume 2, 691. Letter of August 30, 1850.
[21]*Miscellaneous Letters to Surveyor General*, Volume 2: 1848-1856. 1106. Land Records and Title Section, Division of State Lands, Department of Environmental Protection, Tallahassee, Florida. Hereafter, *MLSG*.

legislature is meeting unless you are free from obligations, that is not sitting in the legislature.

His stay in Tallahassee cut short for want of ready employment, it appears that John Westcott returned to Madison and practiced medicine. However, he did not remain tied to his office while there and continued to "dabble in mechanics" and other economic ventures. At about this time, Westcott is recorded as having begun a sawmill on the east side of Range Street in Madison.[22] He also continued to employ his time in community projects and other involvements. One such involvement actually proved a bit embarrassing when he injured himself slightly by misfiring the cannon during the July 4,1851, celebration.[23] His time away from the spotlight was to be short-lived however, and in early 1853 Westcott was named Surveyor General of Florida by President Franklin Pierce.

His tenure as Surveyor General of Florida was to last five very busy years during which time he worked with his deputy surveyors on the surveying of the swamp and overflowed lands, the establishment of the Internal Improvement Trust Fund, the carrying on of the statewide surveys of public lands and cooperated with the U. S. Army in its preparations for the Third Seminole War, which "broke out" with the attack on Lieutenant Hartsuff's command on December 20, 1855. The Third Seminole War, and the events leading up to it, put tremendous pressure on the Surveyor General. The United States policy, from the passage of the Armed Occupation Act of 1842 through the ordering of surveyors into the twenty mile buffer zone, was one of gradual, but continual, pressure on the Indians to live up to their bargain and leave Florida. The native Floridians did not agree with this interpretation of past treaties and resisted the pressure of white settlement as best they could. Because one of the spearheads of this gradual pressure was the official surveying of the lands inside the twenty-mile buffer zone, pressure was brought to bear on Westcott to get his deputies into the area. Westcott, who agreed with the overall goals of this policy, attempted to accommodate the policy at nearly every turn. He was personally convinced that the policy of gradual pressure would get the desired results. The only problem for Westcott in this scenario was the timing of the removal. He was convinced it would happen in late 1853 or in 1854. The trials began in earnest on March 23, 1854, when the Secre-

---

[22]Sims, 37.
[23]Sims, 57.

tary of War, Jefferson Davis, ordered the surveying of the Everglades region to begin.[24]

In early 1854, Commissioner of the General Land Office, John Wilson, wrote to Westcott asking which lands he would recommend for surveying first in the disputed area. Westcott replied that, "The lands on Okeechobee, Kissimmee river and Peas Creek would command the immediate attention of the settlers & larger purchasers, when all fears are quieted of molestation by the Indians."[25] He sent three of his deputies into the area, conscience of the possibility that an incident could, at any time, spark a new war. His men were also aware of the danger, but, undaunted, proceeded to their work. U. S. Deputy Surveyor, John Jackson, acknowledged that the prevailing opinion in Tampa was the possible loss of scalp and life if he and his crew were to proceed to the survey of the Peace River country. After a lengthy silence and no communications to the outside world, Jackson wrote to Westcott on January 12, 1855: "I presume on account of my long silence that you begin to think by this time (with others of our neighbours) that King Billy has got hold of us but such is not the case as you will presently see on my reporting progress."[26] However, Jackson did realize that he was being closely watched and followed nearly every step of the way: "The Indians were watching our movements, even after our crossing Charliepopka Creek and perticularly about the Big Prairie and thence to Istockpoga Lake they set the woods on fire about us frequently; I presume they thought to frighten us from going further on their boundaries...They have been complaining to Capt Casey that we frequently crossed their lines."[27] The U. S. Deputy Surveyors did not present as large a threat to the Indians as did the military reconnaissance parties, who were also actively map-

---

[24]Joe Knetsch, "John Westcott and the Coming of the Third Seminole War: Perspectives from Within," Unpublished paper presented at the Annual Meeting of the Florida Historical Society, May 12, 1990, Tampa, Florida. In this paper, I covered much of this territory and presented the documents upon which this discussion is based. Copies of the paper are on file at P. K. Yonge Library of Florida History, Gainesville, Florida; the Special Collections Section of the University of South Florida Library, Tampa, Florida; and at the State Library of Florida, Dodd Room of Florida History, Tallahassee, Florida.

[25]*Letters of Surveyor General 1853-1860*, Volume 9, 115. Land Records and Title Section, Division of State Lands, Department of Environmental Protection, Tallahassee, Florida. Hereafter, *Letters of Surveyor General*.

[26]*Letters and Reports to Surveyor General*, Volume 2, 152. Letter of January 12, 1855. For additional information on the life and career of John Jackson see, Joe Knetsch, "A Surveyor's Life: John Jackson in South Florida," *Sunland Tribune*, 18 (November 1992), 3-8.

[27]*Letters and Reports to Surveyor General*, Volume 2, 153-54. Letter of February 7, 1855.

ping the territory for totally different purposes. The three parties remained in the field until after the outbreak of the conflict without loss of life or serious injury.

These surveys were valuable to the military and Captain John Casey, the former Indian agent and one of the few men trusted by the Seminoles and Miccosukees, wrote frequently to Westcott requesting maps of the most recent surveys of southern Florida. On two separate occasions prior to the outbreak, he provided the needed maps to Casey and Colonel J. Monroe, then commanding in Florida at Tampa.[28] After the commencement of hostilities, Westcott was very cooperative in providing the most current information gathered from the surveyors to the Army's top topographical experts, including Lieutenant Joseph Ives whose famous map of Florida in 1856 was compiled from the surveyors materials and the reconnaissance reports of his fellow officers in the field. In one of the more telling letters reflecting on the role of the Surveyor General in the conduct of the war, Westcott wrote:

> I have caused to be constructed a diagram of part of South Florida, from the Surveys as made by this office, and the Sketch you were so kind as to present me, of that portion of the Country south of the Caloosahatchee, which I send you. I also send you a copy of letter I received to day from deputy surveyor Wm. S. Harris. There is no doubt but that Indians are in that section of country. Mr. Harris had not heard of the out break when he wrote the letter...I would respectfully suggest that you send at least a Lieutenants command east of the Kissimmee to scout that portion of country from Mr. Harris locality south to Okeechobee.[29]

Not surprisingly, Westcott, himself, was in the field on an inspection when news of the outbreak of hostilities was relayed to him at Fort Kissimmee.[30] Unlike previous Surveyor Generals of Florida, Westcott was often in the field to assure Washington and himself that the surveys were being carried on correctly, which also increased his personal knowledge of the territory and exposed him to the same dangers faced by his deputies, but his information to Casey, Monroe and Ives was more correct than any that could have been provided by his predecessors.

The Third Seminole War took up the remainder of his tenure as Surveyor General of Florida and limited his men to surveys of grants, donations and other

---

[28] *MLSG*, Volume 2, 271. Letter of June 7, 1854 and Letter of April 25, 1855.
[29] *Letters of Surveyor General*, Volume 9, 269-70. Letter of January 17, 1856.
[30] *Letters of Surveyor General*, Volume 9, 270-71. Letter of January 18, 1856.

"scrap" work left undone in relatively safe locations. By mid 1858, John Westcott was looking for new employment, but not done with the controversy involving his tenure as Surveyor General. In June of that year, he began a rather acrimonious correspondence with the new Surveyor General, Francis L. Dancy, over his living quarters in the barracks at St. Augustine. Westcott maintained that Quartermaster General, Thomas Jesup, had agreed to allow him to keep his private rooms in the barracks and that they were *not* part of the rooms granted for the use of the Surveyor General's offices. Westcott noted to Dancy that the barracks had not been fully occupied for some years and that Jesup preferred to have all of the rooms in use and maintained by the occupants, which Westcott did, at his own expense. Dancy, of course, believed that these rooms were part of those he obtained for use in his official duties.[31] This rather trivial squabble was more than a quest for more office space on the part of Dancy, it appears to be a portion of a plan to get at Westcott for political purposes, as the former Surveyor General was now a candidate for Congress as an Independent Democrat (Know-nothing).

Westcott's political gamble was a weak one from the start. Although well known throughout the state and respected for his learning, his choice of political "party" left much to be desired. Even before he officially declared for the office of Congressman, he was castigated by the *Floridian & Journal*, the Tallahassee based organ of the state Democratic party, in the following words:

> We alluded last week to a political demagogue calling himself a Democrat, who is represented to us as being busy in efforts to induce divisions in the Democratic ranks in counties East of Tallahassee. We suppose our readers understood to whom we referred. He is the Ex-Surveyor General of Florida. His expulsion from office is the cause of his attempt to give to the contemplated Alligator meeting in imposing character. We stated a week ago, that that meeting was of Know Nothing origin, but we now have reason to believe that *Westcott* has become its chief director. It is an office precisely suited to his genius. Wanting in that kind of ability that commands respect, but not without a certain degree of smartness, he makes up all deficiencies by low unscrupulous cunning.[32]

This was the mild beginning of the opinion that painted only the darkest of pictures of the "Ex-Surveyor General." Within less than a week, Westcott had

---

[31]*MLSG*, Volume 3, 161-71. Letter of June 7 and two of June 8, 1858.

[32]*Floridian & Journal*, July 24, 1858.

announced his official candidacy against the incumbent Democratic Congressman, George S. Hawkins, who had gone into office in the Democratic sweep of 1856. Westcott did not hide the fact that he looked to support to old Whigs, Know Nothings and disaffected Democrats, and committed himself to backing the unionist policies of the Buchanan administration. One of his major planks, as he had argued while in the legislature and in an 1856 election pamphlet, was a low, affordable price for state lands. He declared himself a follower of the principals of Andrew Jackson and openly avowed his first vote was for the General and his last for Buchanan. In some of his best campaign rhetoric, he stated:

> In advocating cheap lands - opposition to monopolies - free agency in voting, and the right of citizens to become candidates, independent of the caucuses, or of conventional nominations, I violate no *principals* [*sic*] of teachings of the Fathers. I must confess, however, I never knew a part so just, so reasonable, and strictly right in *all* things that a man could be *tied* to it, strictly and blindly follow it, in its height and violennce, and at the same time to observe the divine injunction, *do right*. Fidelity to the true interests of the State is a higher and more imperative obligation than that to party organization, especially when used by a *few* improperly.[33]

Acceptance of the Buchanan administrations interpretation of the fugitive slave law, its weak stand on territorial self-rule, and other policies did not give Westcott a strong platform to run on in a state rapidly moving toward succession and being opposed by the regular Democratic machine, including both Senators and Congressman Hawkins. Even strongholds of leftover American Party (Know Nothing) sentiment, such as Hillsborough County, did not give Westcott consistent support.[34] The end result was very predictable, an election loss by over two thousand votes.

John Westcott had other distractions in the year of 1858. The State Legislature, in its early session of that year, granted a charter to the St. Johns Railroad Company. The Board of Directors of this new venture to construct a railroad between

---

[33] *Floridian & Journal*, July 31, 1858.
[34] A good discussion of the decline of the Know Nothing movement in Hillsborough County can be found in Spessard Stone, "The Know Nothings of Hillsborough County." *Sunland Tribune*, 19 (November 1993), 3-8. A brief discussion of Westcott's election can be found in Arthur W. Thompson, "Political Nativism in Florida, 1848-1860: A Phase of Anti-Successionism," *The Journal of Southern History*, 15 (February 1949), 60.

St. Augustine and the St. Johns River, were John Westcott, Richard F. Floyd, B. E. Carr, D. G. Livingston and C. Bravo, with Floyd being elected the first president of the line. According to the newspaper account, "The Directors have assured us that in less than twelve months we shall ride in comfortable cars, after the "iron horse," to the river in thirty minutes."[35] The optimism which met the roads organization was felt throughout the area and great things were looked for from the new line of communication. These expectations were not to be met until many years after the Civil War, when new investors and more funds made getting an iron horse a reality and not a dream.

Westcott had not only invested in the railroad, but obtained valuable experience in the operation and construction of railroads when he was appointed examining engineer for the Florida, Atlantic and Gulf Coast Railroad by the Board of Trustees of the Internal Improvement Trust Fund in July of 1858. As such, he investigated the grading, laying of the ties, and other details of the construction of the line as were in compliance with the orders of the Trustees.[36] This experience must have been of some value, as this line was more ambitious and complex than that proposed by Westcott, Floyd and partners.

As a practical surveyor and prudent investor, John Westcott laid out the initial route for the new railroad. He surveyed every inch of the way and noted the major obstacles that would be encountered. The route lay between Tocoi Landing on the St. Johns River and to the west bank of the San Sebastian River on the outskirts of St. Augustine. According to Greville Bathe, railroad historian, the reason for stopping short of the city proper was the large amount of intervening marsh along the river that would have to be crossed, a very expensive proposition.[37] As Bathe described it: "The original tracks of the St. Johns Railroad were simply squared wooden stringers, laid on the felled trees cut down during the clearing of the right of way, and were protected on the top by thin strips of iron, similar to that used on tires of waggon wheels, and around 1858, sold for twenty dollars a ton."[38] This description is similar to other early railroads of Florida, especially the Tallahassee to St. Marks railroad. The early cars, once the construction was completed for the first twelve miles, were drawn by horses or mules, which meant that two animals

---

[35] *Floridian & Journal*, February 2, 1859.
[36] *Board of Trustees of the Internal Improvement Trust Fund: Minutes,* Volume 1 (Tallahassee: L. B. Hilson, 1902), 91.
[37] Bathe, 2.
[38] Bathe, 2-3.

were used to haul the freight over a broad-gauged rail, i.e. five feet apart. This arrangement was not rapid nor satisfactory as the mules often balked at their task. In 1860, Westcott and friends were forced to rethink and redo most of the roadbed to accommodate a new mode of power, the iron-horse. However, as reported in the newspapers, by May of 1861, although the road had been regraded to within one and a half miles of St. Augustine and rails laid over the first mile and one half, construction had to be stopped because of "political troubles" then brewing, namely the firing on Fort Sumter and the beginning of the Civil War.[39]

Westcott, although a man of some unionist principals as noted in his defense of Buchanan's policies, did not let the coming of the war interfere with his proposed projects. In April of 1861, while work was continuing on the railroad, he proposed to the Board of Trustees the possiblity of draining swamp and overflowed lands in the reaches of the upper St. Johns River for the growth of sugar and other products. The Trustees, responding to this proposal, agreed to allow Westcott to survey the area, run the necessary levels to ascertain the practicability of such a project, and allowed themselves latitude to determine the compensation based on the results of Westcott's survey. The area to be surveyed for improvement and drainage lay south of Township 20, South, or in the area of Salt Lake, near Titusville, down to about the vicinity of Lake Poinsett, across the way for modern Rockledge. Not coincidentally, this was the area near the route of the St. Johns and Indian River Canal Company, headed by J. G. Speer.[40] His interest in this area was to continue after the war.

Just prior to the war, Westcott was active in a number of local concerns. In April of 1860, he was elected as High Potentate of the Florida Royal Arch Chapter No. 2.[41] Earlier that year, he presented a paper of significance before another relatively new organization, the Florida Historical Society. In this groundbreaking paper, he discussed, based upon his surveying experience, the probable route DeSoto took across Florida. The newspaper of the time stated: "We cannot but regard his address, as a most valuable communication on this subject, and one which will do much towards it, if it does not entirely settle this disputed route."[42] Although this did not end the dispute, as Buckingham Smith returned a few weeks later with new information from the Spanish Archives in Seville which differed in interpre-

---

[39]Bathe, 4-5.
[40]*Trustees Minutes*, Volume 1, 215-226.
[41]St. Augustine *Examiner*, April 7, 1860.
[42]St. Augustine *Examiner*, February 4, 1860.

tation from that presented by Westcott, it did add more fuel to the debate which rages on today.

The war ended his social and economic ventures for some years. His record in the war shows an older soldier who was willing to take command and contribute. The official records show that John Westcott was mustered into service in 1862, with no specific date listed. It is well known that he served at Tampa early in the war and was present at the Battle of Olustee as an officer in the 6th Florida Battalion. Exactly what his assignment was is unclear, the picture of him in his uniform at the Confederate Memorial Literary Society, in Richmond, Virginia, notes his position as surgeon, 6th Florida Regiment. The official records list him as a major in the 10th Florida Infantry, 2nd Battalion. Dickison's *Military History of Florida* in the *Confederate Military History* series states that immediately following the Battle of Olustee, Westcott's company was ordered to the Ocklawaha area to assist in heading off an attempted raid in that sector.[43] Therefore, there is some confusion as to Westcott's service in Florida during the war. What is equally clear from the records, he served in the Virginia campaigns of later 1864, including Petersburg and Cold Harbor, and mustered out of service on April 9, 1865.[44] The date of his mustering out is significant in that Westcott, along with the remainder of the 10th Florida Infantry, was present to surrender at Appomattox Court House. He was one of eighteen officers and one hundred and fifty-four men who survived the rigors of war.[45]

Westcott's return home was not a joyful one, like most who came back to Florida. His immediate dream of revitalizing the railroad was not to be, for, as he wrote, "During the War, the depots were burnt, the rolling stock destroyed, nearly all the iron carried off, leaving only the road bed, land, & franchise to the owners. Since, for want of <u>ready means</u> we have been unable to re-construct it."[46] The exact status of the road until 1870 is another puzzle in that Bathe represents that a small "Coffee Mill" engine was running on the line between Tocoi and St. Augus-

---

[43] J. J. Dickison, *Military History of Florida*, Part of *Confederate Military History*, Volume 1, (Secaucus, New Jersey: Blue & Grey Press, n.d.), 83.

[44] Florida Department of Military Affairs. "Florida Soldiers: CSA 9th, 10th and 11th Infantry," Special Archives Publication No. 93. 219 and 233. These records show him to be part of the field staff of Colonel Charles F. Hopkins, a fellow surveyor, and as a Captain of I Company. His promotion to Major came after his immediate service in Florida and probably came with the commands transfer to Virginia, or shortly thereafter.

[45] "Florida Soldiers: CSA," 219.

[46] Florida Department of State, Archives and Record Management. Series 914, Box 14. Letter of March 15, 1870. John Westcott to Board of Trustees of the Internal Improvement Trust Fund.

tine in the 1866 to 1870 period. He goes to great length to quote from Charles Hallock, noted author and editor, and Dr. Andrew Anderson, the prominent St. Augustine physician and Westcott's friend, about the state of the railroad in 1868. However, the petition quoted above was written in March of 1870 and indicates that the road was not operating its full length, fourteen and one-half miles. This is reinforced by the fact the Board of Trustees of the Internal Improvement Trust Fund passed a resolution on behalf of Westcott and the railroad granting additional lands for the "completion of the Road and its necessary drains and ditches," at which time the title to the granted lands would be given to the railroad after a certificate of completion had been issued by a competent engineer.[47] Additionally, most of the correspondence found for Westcott during the period between 1866 and 1870, concerns his law and land practice out of Jacksonville and not the matters of the railroad.[48] The problem presented here may not be answerable until further correspondence emerges in the future, however, the one thing that is certain, is the fact that all of Florida suffered from the war and Westcott and his partners in the railroad were not immune.

And although John Westcott eventually sold his interest in the St. Johns Railroad to William Astor and others, he remained on the Board of Directors until he died, demonstrating a commitment to the enterprise until the very end.

John Westcott also had other dreams regarding internal improvements. Another type of railway was envisioned by the inventor, lawyer, Surveyor General was a "saddlebag" railroad. This unique invention, which he patented, consisted of a single rail system that is similar in concept to today's monorail system. He demonstrated his creation at the famed Centennial exposition in 1876.[49] The energetic inventor soon interested some prominent men of Florida to back him in forming the "No Gauge or Single Rail Railroad and Construction Company" in early 1876 and got the legislature to incorporate it. Westcott, on behalf of his new company, then asked the Board of Trustees of the Internal Improvement Fund to sell swamp and overflowed lands cheaply to the company along a route from

---

[47] *Trustees Minutes*, Volume 1, 416-17.

[48] Letters found to date include ones of March 28, 1866, October 19, 1866 and February 2, 1870. Florida Department of State: Archives and Record Management. Series 914, Boxes 12 and 14. Other letters found in the *MLSG 1869-74*, Volume 4, 9, 20, 34, 57, 100 also indicate his major concern was the law and land practice. All of these, however, do not convince this researcher that one can assume that the above statement is true. Too much of Westcott's correspondence is missing to preclude that an opposite conclusion could not be reached.

[49] Rerick, 158.

Orange Lake to the Ocklawaha River in Marion County. The Trustees reserved lands along the route for six months, at which time the road was to have been completed.[50] The road was not built or even graded. It appears that the time limit was too short and the capital needed was not forthcoming, therefore, the scheme had to be abandoned. But this minor setback did not dampen the spirits of the good Doctor and he went on to his next, and most important, enterprise.

The dream of an intra-coastal canal extending from the St. Augustine to Biscayne Bay was not original with John Westcott. Many men before him had contemplated the idea, including the Spanish who had used the natural "inside passage" during the first Spanish occupation. With the beginnings of the Internal Improvement Board, the predecessor to the Board of Trustees of the Internal Improvement Fund, member John Darling of Tampa had solicited ideas concerning such a canal route and found a favorable response from the Miami postmaster, George W. Ferguson.[51] Westcott himself noted to Issac Coryell that he had contemplated the canal as early as the organization of the St. Johns Railroad and considered it "part of the projected work" that would link transportation systems and tap the unused resources of southern Florida.[52] The idea, however unoriginal, had not been attempted with any serious funding, unless one includes the short-lived attempt by William Gleason in 1869-72, and it is Westcott who is most often credited with the title, "father of the East coast system of canals."[53]

Many things delayed the beginning of the Florida Coast Line Canal and Transportation Company's work in constructing the intra-coastal canal. There were many competing claims for land, such as those of the St. Johns and Indian River Canal Company, the Atlantic Coast Steamboat, Canal and Improvement Company and the old Hunt and Gleason scheme. Added to these difficulties was, of course, the famous litigation surrounding the actions of the Board of Trustees of the Internal Improvement Trust Fund and one Francis Vose. The Vose suit and injunction delayed the construction of most internal improvements in Florida for nearly a decade. Once these legal problems were surmounted, the technical difficulties

---

[50] *Trustees Minutes*, Volume 2, 136-37.

[51] Joe Knetsch and Paul George, "Life on the Miami Frontier," *South Florida History Magazine*, Fall 1990, 8-9. The article is an edited version of the original letters found at the Department of Environmental Protection's Land Records and Title Section in a rectangular file box labeled, "Swamp and Overflowed Lands."

[52] Letter of January 14, 1882. Westcott to Coryell. Photocopy of the original letter is located in the State Library of Florida's Special Collections (Dodd Room) at the Florida Department of State, Tallahassee, Florida.

[53] Rerick, 158.

were encountered. This complicated project took a great deal of engineering skill, funding and political cajoling to complete. The initial surveying of the route was taken on by Westcott himself, though he had little practical training in this type of civil engineering and was later, after his death, criticized for his planned route. The project was aided by earlier construction over difficult places, such as the famed "Haulover Canal" area, by the Army Corps of Engineers.[54] However, most of the early work was done by the company without any outside assistance except in the form of land grants from the Board of Trustees of the Internal Improvement Fund. Although the company was incorporated in 1881, it did not have the funds or equipment to begin work until 1883. As Westcott wrote to Coryell in January of 1882:

> This canal is an immediate necessity. All the Settlers South of here on the Coast, are clamerous for the work to be commenced at once. The Halifax country is now an important point to Secure, or other improvements may be made. & so it is with the Indian River Country. The shipments of fruit & vegetables are now large, & under the most adverse Circumstances, and so with their receipts of Merchandise. If this Canal was now finished & which must be accomplished for the next crop, and for passengers to easily get South I believe it would not be a difficult matter to put from three to five thousand Settlers, on the line of improvement in a Short time. No man now can Settle on the coast water line, because what he produces is locked up, for want of quick & Certain transportation.[55]

Westcott's sense of emergency was real enough, but it could not get the canal built in time for the shipments of 1883. Indeed, not until more funds and equipment were available could the canal become much of a reality at all.

The original requirements for the canal were fifty feet in width and five feet in depth, enough to float steamers that normally were plying the trade of the Ocklawaha and upper St. Johns above Lake Monroe. The originally constructed canal, however, was a bit smaller than the required depth and width, primarily so that these types of craft could get the goods to and from market sooner. The result was, as Westcott admitted to Coryell, a short term projected canal of only thirty

---

[54]George E. Buker, *Sun, Sand and Water: A History of the Jacksonville District Army Corps of Engineers, 1821-1975* (Jacksonville: U.S Army Engineer District, 1981), 116-17.

[55]Westcott to Coryell, letter of January 14, 1882.

feet in width and three feet in depth. Westcott, in his optimism for the project, was very correct in predicting some of the results of this canal, even if the requirements were not strictly met at first. One of his predictions was the growth of a new city on Biscayne Bay, or Barnes Sound, which would soon eclipse Key West and open a new line for commercial products, primarily from tropical fruits.[56] Although this prediction became true, Westcott did not live to see it even begin nor did he last long enough to see the canal reach the required depth and width.

After turning over the actual construction of the line to the construction crews, Westcott spent the remaining years arguing before the Board of Trustees of the Internal Improvement Fund for grants of land or proposing other grants in lieu of those that could not be fulfilled. As late as April 26, 1888, Dr. Westcott, as he was always referred to in the official documents, was stating the canal company's case before the Board of Trustees.[57] The constant travel and mental strain of these trips to Tallahassee must have taken their toll on the elderly doctor. In the long run, however, his efforts helped to obtain 1,030,128 acres of land for the company which, in turn, paid for the construction of the intracoastal canal, which was not finished until 1916.[58]

The efforts of this man's remarkable life brought many changes to the lives of all Floridians. Anyone who travels the intracoastal waterways owes a bit of gratitude to John Westcott. Anyone who travels comfortably or ships their fruits over the railroads of Florida can look to John Westcott with a knowing eye that his efforts helped to bring Florida into the railroad age. And anyone who owns property in the Sunshine State must realize that as a surveyor and as Surveyor General, John Westcott had something to do with maintaining the integrity of their property lines. Doctor Westcott's life, in Florida, was one dedicated to the public welfare, from nursing the wounded in two wars to initiating the construction of the intracoastal canal system, his was a life of service. Like all of us, he made his enemies and paid for his mistakes. In the end, however, this neatly trimmed man in the white calico suits with soft mustache and goatee beard, must stand above most others as a leading pathfinder of Florida and a precursor to the modern world.[59]

---

[56]Westcott to Coryell, letter of January 14, 1882.
[57]*Trustees Minutes*, Volume 3, 489-90.
[58]Buker, 117.
[59]This physical description is taken from Bathe, 10.

CHAPTER 4

UNTIRING, FAITHFUL AND EFFICIENT: THE LIFE OF FRANCIS LITTLEBERRY DANCY

One of the ironies of reading history is the fact that men and women who were very important to their contemporaries are often forgotten or overlooked by historians. In Florida, individuals like Odette Philippe, Benjamin Putnam, Issac Bronson, William Cooley, John Darling and Francis Littlebury Dancy are excellent examples of this type. All of these men fought in the Indian Wars, started communities, held important offices and helped to shape the frontier of the State of Florida, yet, without exception, none appear in the general history books of Florida. The major reason or excuse given is that their papers are too hard to locate and are not readily available. But this does not preclude the existence of such papers or information related to them. Diligent research will locate enough data on the lives of these individuals to enable a capable historian to recreate their lives in some detail and demonstrate that the world did not revolve around the great and near-great men of Florida's past. Indeed, without an understanding of these people, the lives and events surrounding the "great men" are almost meaningless and devoid of depth.

The life of Francis Littlebury Dancy was important to the history of Florida's development. As an engineer, he opened up the frontier by clearing the Ocklawaha River, opening a major road inland to Fort King and constructing the famous seawall in St. Augustine. As a military leader, he led troops during the Second Seminole War and helped to organize the forces at the onset of the War Between the States. As a surveyor and Surveyor General of Florida, he helped to organize the property lines of the state and opened up large areas of land for settlement. As a political leader, he served as a mayor and alderman in St. Augustine, one term in the legislature as a representative and was the acknowledged leader of the Democratic Party in St. Johns County and, later, Putnam County. As a horticulturalist, his development of the Dancy Tangerine and the magnificent groves at Buena Vista gave the Dancy name worldwide recognition and encouraged others to develop Florida's citrus potential. Such a life deserves recognition.

Much of the early genealogy of the Dancy family has been done and is now located in the Putnam County Archives, however, a brief review of the information will give some of the necessary background. F. L. Dancy was born in Edgecombe County, North Carolina, in 1806, the son of Edwin Dancy and Lucy Knight. His grandparents were William Dancy of Edgecombe County and Agnes (or Agatha) Littlebury. The great-grandfather was also named William Dancy but he was from Sussex County, Virginia and was married to Mary Mason of Albemarle Parrish in Sussex County.[1] This is the family line from which F. L. Dancy descended.

Francis Littlebury Dancy received his early education at home, probably from tutors, and entered the United States Military Academy at West Point on July 1, 1821. Following the usual curriculum of the day, Dancy studied mathematics, surveying, the physical sciences and history. He graduated and was entered the United States Army on July 1, 1826, and was sent to the Artillery School at Fort Monroe, Virginia, until 1828.[2] He then served improving the inlet at Ocracock, North Carolina, until the next year, when he was assigned duty in the Topographical Engineers under Major J. D. Graham surveying the route for a canal through South Carolina. He served as a surveyor under Colonel Stephen Long for a turnpike through Eastern Kentucky, Virginia, Eastern Tennessee and along the Blue Ridge into North Carolina. Dancy also saw duty at Muscle Shoals, Alabama, improving the road between there and Knoxville, Tennessee, in 1830-31. The following year, he was transferred with his regiment of the 2nd Artillery to Fort Moultrie, South Carolina.[3] In late 1832, Dancy received his commission to as First Lieutenant of the 2nd Artillery. By 1833, his regiment had been transferred to St. Augustine, Florida, and he was assigned the supervision of the repair of the sea wall and the walls surrounding Fort Marion (as the Castillo was then called).

---

[1]"The Family of Francis Littlebury Dancy." From the files of the Putnam County, Florida, Archives, Palatka, Florida. Author and date of writing unknown.

[2]George W. Cullum, *Biographical Register of the Officers and Graduates of the U. S. Military Academy*, Volume 1 (Boston: Houghton, Mifflin and Company, 1891), 369.

[3]"Department of the Interior Appointment Papers: Florida, 1849-1907. Roll 1: Surveyor General [A-D], 1849-1907." Washington: National Archives, 1980. Microfilm M1119. Letter of June 4, 1877. Dancy to Carl Schurz.

In 1835, he was given the additional responsibility of repairing the road between St. Augustine and Pensacola, known popularly as the Bellamy Road.[4]

While on duty at St. Augustine, Dancy met and fell in love with the daughter of Judge [later Governor] Robert Raymond Reid. Francis and Florida Reid were married before the end of 1833. This happy marriage lasted well past their golden anniversary.[5] Reid was a very astute politician and was well connected to the existing Democratic power structure of East Florida. Dancy, whose views on politics often mirrored his father-in-law's, was an eager student of Territorial politics and had had an insiders view of the very political nature of the United States military. He was soon a favorite officer in the command of General Duncan L. Clinch, also a strong Democrat and a rival of General Winfield Scott, later a candidate of the Whig party for president of the United States. It was Clinch who recommended Dancy for the post of Assistant Quartermaster of the United States Army in Florida after the outbreak of the Seminole War in December 1835.[6] The union of love and politics was to have a permanent affect on the life of F. L. Dancy.

Dancy began his career in Florida with the work of repairing the Bellamy Road between St. Augustine and Tallahassee. But, on April 7, 1834, a large group of inhabitants of St. Augustine petitioned the government to repair the sea wall and the ancient fortress. As these men noted, "That the encroachment of the waters into the harbor and upon the city have destroyed many buildings and nearly the whole of one street." The old Spanish fort, then called Fort Marion, was greatly endangered and these citizens recognized the value of its preservation, not only for its historical significance, which they fully understood and appreciated, but as an arsenal and store house for the United States Army. This was one of the first calls for historic preservation in Florida history. They also requested the government to extend the appropriations further and to finish construction of the sea wall extension as provided for in an earlier plan.[7] As a young engineer willing to take on difficult tasks, Dancy was charged with supervision of the work. Unfortunately, the appropriation was too small to adequately complete the work.

---

[4]Dancy Letter, June 4, 1877. Dancy to Carl Schurz.

[5]"The Family of Francis Littlebury Dancy." Putnam County Archives. Also see Sarah Margaret Kaiser, *My Family* (Hastings, Florida, n.d.), in the Putnam County Archives.

[6]"Letters Received by the Office of the Adjutant General (Main Series) 1822-1860. Roll 122. D 90-E, 1836. Washington: National Archives, 1964. Microcopy No. 567. Letter of June 27, 1836. Clinch to Brigadier General R. Jones.

[7]Clarence Edwin Carter, Editor, *The Territorial Papers of the United States*, Volume XIV, *The Territory of Florida, 1828-1834* (Washington: National Archives and Records Service, 1959), 997-99. [Hereafter, *Territorial Papers*, volume and page number.]

In March 1835, Dancy was assigned the task of clearing the Ocklawaha River to make it more convenient to ship troops and materials to Fort King and the nearby Indian Agency, just three miles from the headwaters of the famed Silver Springs, a tributary to the Ocklawaha. In accordance with his orders, Dancy consulted with General Clinch and began the operations in November 1835. During this same time period, he supervised the repair of the road between St. Augustine and Picolata, which provided the most efficient means of resupplying St. Augustine via the St. Johns River, should the sea routes be endangered or the bar closed. Lieutenant Dancy was very concerned about the coming operations in his theatre and recommended that this road be quickly repaired as part of the appropriation for the fixing of the Bellamy Road. Additionally, he strongly urged the use of the settlement at Palatka as the jumping off point for any reinforcement or supply of interior posts. Specifically, he noted the road to Picolata as important as the first step in transporting troops and supplies to Palatka and then going inland over the road to Fort King. He was asked to repair this latter route, too.[8] The Army was fortunate because Dancy, at the commencement of hostilities with the Seminoles, immediately halted the shipment of materials to Palatka and order them stored at Picolata. The reason for this action was the burning of the little settlement in December 1835.[9]

With the commencement of the war, Dancy joined his regiment and rushed to the interior to aid General Clinch. He took part in a series of scouting operations leading up to the Battle of the Withlacoochee. He was stationed at Clinch's Auld Lang Syne Plantation, the site of Fort Drane, and was the officer in charge of the post when the famed battle took place. He was in charge of the sick, the medical corps and a few officers and enlisted men, not totaling fifty men. Using his engineering skills, Dancy immediately set about constructing breastworks and two small blockhouses, dubbed "Camp Dancy" by those in attendance. When a group of Florida militia came to the camp and demanded rations, Dancy, who suspected the men were deserters, refused and ordered his little garrison to stand at attention and prevent any attempt at seizure by this group. Shortly after this group left, a single soldier came riding up to the compound and notified Dancy of the battle

---

[8]*Territorial Papers*, Volume XXV, 112, 117-18, 136, 163-64, 232-33.

[9]*Territorial Papers*, Volume XXV, 232-33. Dancy noted that he had three lighters at Palatka at the time of its capture and burning. He assumed that they were lost.

and the immediate need of ammunition. Dr. John Bemrose and others immediately jumped to the ammunition wagons and began loading the rider down with all that he could carry. Dancy then ordered fires started about one hundred yards from the pickets to give light in the darkness and keep the enemy at bay. Every man capable of bearing arms was put in the rotation guarding the newly created facility. After news of the battle, Dancy was sent forward to retrieve the baggage train and to secure it at Fort Drane. The scene of the returning wounded was pathetic and many of the wounded died at that place. The fortification was considered unhealthy by the medical staff, who urged its abandonment from May 1836, onward.[10]

Dancy, was assigned garrison duty at Fort Drane, which he noted was, "considered a perfect grave yard."[11] Because he felt his health slipping and he had been offered more gainful employment, Dancy resigned from the United States Army in a letter of July 22, 1836. His health and the future of his young family weighed heavily on his mind. And, since his regiment had not been relieved of interior duty during the hot, unhealthy summer months, he believed he had little choice but to leave the service. In this letter, he correctly noted that nothing would or could be done in those hot months during the rainy season. This observation was to remain true throughout this and the Third Seminole War (1855-58).[12]

Almost immediately upon his resignation, F. L. Dancy was given the contract to finish the work on the sea wall and the repairs at Fort Marion. The contract was for fifty thousand dollars and Dancy was to receive three dollars per day and two and one half percent of the amount disbursed.[13] The former Army engineer faced some formidable problems, not the least was the shortage of skilled labor. He also faced the enmity of his political enemies, many inherited from his alliance with Robert Raymond Reid. Of this group, the most powerful was Florida's Congressional Delegate, Charles Downing and his allies at the St. Augustine *News*. Almost from the first, the appointment of Dancy as the contractor was questioned in the most partisan way. After nearly two years of work on the project, Dancy's opponents succeeded in wresting the project away from him and getting it placed

---

[10] John Mahon, editor, *Reminiscences of the Second Seminole War by John Bemrose* (Gainesville: University of Florida Press, 1966), 43-55.

[11] Letters Received by the Office of the Adjutant General (Main Series) 1822-1860. Roll 122. Letter of July 22, 1836. Dancy to General R. Jones. Dancy underlined this phrase in his letter.

[12] *Ibid*

[13] *Territorial Papers*, Volume XXV, 327-28.

into the hands of a Lieutenant Benham, an ally of Downing. Dancy's response was to ask his former fellow officers for an official hearing.[14]

The main charges leveled against Dancy by Benham and Downing were that of malfeasance in office and profiteering. They also accused the him of using laborers on his personal projects, particularly the St. Augustine Heights and at Shell Bluff. Benham claimed that Dancy refused to turn over all of the Government's papers regarding the construction of the sea wall and the repair of the fort, offering only to copy said documents at his leisure. Benham charged that this was too time consuming and delayed the project, since he needed many of these documents to make out his own reports and get acceptable models for the filing of vouchers and other paperwork. Dancy, of course, refused to acknowledge the validity of the charges, noting that he no longer had an assistant to copy the voluminous reports and had to do all of this without aid. He did acknowledge the use of masons, carpenters and their assistants on his personal projects, but argued that they were paid from his own pocket and were never charged to the United States. Indeed, he maintained that these men were used at slow periods to help secure them a more steady income and employment. Also, he noted, these men were the only skilled laborers available for just about any project, therefore he had little choice but to hire them when his personal needs required.

The testimony in the hearings with Captain J. K. F. Mansfield of the Army Corps of Engineers, took a few weeks to complete. Many witnesses were called and examined by both sides. The results were not what Downing and his allies at the St. Augustine *News* were hoping. Mansfield reported that some irregularities occurred but that none were of a serious nature. Both sides had called their allies to witness and much of the testimony was, plainly, hearsay evidence. The final verdict was published by Secretary of War Joel Poinsett, "The department concurs in the opinion expressed by the Chief Engineer and trusts in the future that it will be distinctly understood that no agent will be permitted, on any pretext whatever, to employ public materials or labor on his private works. It is not deemed expedient, upon a review of this whole case, to direct any legal proceedings against Mr. Dancy."[15]

---

[14]United States Congress, House of Representatives Report No. 201. 26th Congress, 1st Session. May 12, 1840. "Lieutenant F. L. Dancy." This document contains almost all of the correspondence that passed between the combatants in this skirmish of political wills.

[15]*Ibid*, 118. It should be noted that I have summarized 124 pages of material into approximately three single paragraphs. The documents presented make for interesting reading and show the partisan nature of the entire affair.

It was while this investigation was taking place that Dancy was elected to a term as the Mayor of St. Augustine, from January 1838 through November 1840.[16] Dancy's term saw numerous ordinances passed, mostly with the rules governing the old City Market, police and maintaining the peace. During this time, the city passed an ordinance allowing the mayor to appoint a city marshal who reported directly to that officer, while enforcing the laws passed by the mayor and city council. This tightening of the power of the city indicates that some lawlessness was a problem in the Ancient City.[17] This impression was reinforced by the passage of an act prohibiting anyone from appearing in public in a state of intoxication or acting in a disorderly manner.[18] The "Whereas" clause of the enactment stated clearly that the disorderly conduct was because of "the men let loose upon the community by the cessation of operation by the Army." This would also explain why Dancy and the city council passed an ordinance forbidding the opening of liquor stores after 9:00 o'clock in the morning on the Sundays. Something had to be done to protect the families and children from the actions of the drunken soldiers and their hangers-on.[19]

Dancy had a particularly busy year in 1840. In addition to fighting the battles of the City Council, he was in the midst of developing his St. Augustine Heights and Village of St. Sebastian properties. The affair of the sea wall and fort repairs was also winding down to its inglorious end and the Seminoles were still raiding as far north as Mandarin on the St. Johns River, necessitating the calling out of the militia. Once again, Francis L. Dancy answered the call to duty and, as the elected Lieutenant Colonel, and later full Colonel, of the Florida Mounted Volunteers, he led his forces into battle. On September 8, 1840, Dancy and his troops assisted the regular Army in an action near Fort Wacahoota, northwest of Micanopy in Alachua County. The combined force of regulars and militia forced the Indians out of the hammock where they had ambushed a force just south of Fort Walker. Although no Indians were taken or killed by Dancy's small force, he did not hesi-

---

[16]Thomas Graham, *The Awakening of St. Augustine: The Anderson Family and the Oldest City: 1821-1924* (St. Augustine: The St. Augustine Historical Society, 1978), 267.

[17]*Florida Herald and Southern Democrat*, December 24, 1840. The Ordinance was entitled, "An Ordinance regarding the appointment of City Marshall, Clerk of the Market and for other purposes."

[18]From the Dancy files at the Putnam County Archives, Palatka, Florida and compiled by Ms. Nancy Alvers, who took the information from the St. Augustine newspapers for 1838. The ordinance cited here is entitled, "An Ordinance In addition to 'An Ordinance respecting the peace and police of the City of St. Augustine.'"

[19]*Ibid.* The author would like to thank Ms. Alvers and Janice Mahaffey of the Putnam County Archives for the collecting and copying of this material and making it available to researchers.

tate to lead his men into the dense hammock and help drive the enemy from the field.[20] With a growing family, military obligations and campaigns, business opportunities and political controversy, little wonder that 1840 was particularly hectic for Dancy.

Real estate sales in the middle of an Indian war are often not brisk, which sometimes necessitates the search for other means of obtaining wages to help raise a family. In 1842, he called upon his friend and the Territory's Delegate to Congress, David Levy, to assist him in obtaining a contract for surveying the public lands. Levy wrote to the Surveyor General, Valentine Conway of Virginia, on Dancy's behalf. In his letter he declared:

> My principal and first anxiety however is in favor of Col. F. L. Dancy of this place. Col. D. was educated at West Point and continued in the army a long time, most of the while engaged upon Engineers duty. He married a daughter of the late Gov. Reid and afterwards resigned. He is desirous to engage himself in your service and I would be under lasting obligation to you to facilitate his views as far as he can make his offer consistent with the public interests. He is an active, faithful and reliable man, and an experienced and educated mathematician & surveyor....[Levy continued] I think with you that a location here would be most convenient, and I doubt not would be very agreeable to you upon other accounts. I shall leave here about the 1st Novr. for the West, and will see you at Tallahassee, when we can have a full discussion and understanding of the steps you would desire taken at Washington.[21]

Dancy got the position and embarked on his public lands surveying career as the Surveyor of Private Grants in East Florida. He would later repay his friend for his assistance and work with him in the construction of Yulee's dream, the Florida Railroad.

---

[20]Letters Received by the Office of Adjutant General (Main Series) 1822-1860. Washington: National Archives and Records Administration, 1964. Roll 202. Microcopy No. 567. Letter of September 8, 1840. Capt. S. Hawkins to Lt. R. C. Gatlin.

[21]"Applications for Employment. Volume 1: 1824-1844." 261-62. Land Records and Title Section, Division of State Lands, Florida Department of Environmental Protection, Tallahassee, Florida. Conway was very eager to get back to Virginia where family and business interests appears to have interested him more than the Surveyor Generalship of Florida. He is the only such officer to spend more time out of the Territory/State than in it in Florida history. He was later asked for his resignation because of his misuse of public moneys.

But prior to his service in the surveying field, Dancy was elected to the Florida Legislature as the representative from St. Johns County. His assignments in this volatile session included the chairmanship of the Committee on the Militia, and membership on the committees for Corporations and Enrolled Bills.[22] On Friday, January 7th, he introduced a petition on behalf of "sundry Citizens of Duval County" asking for a charter for a ferry across Black Creek at Garey's Ferry, near the site of the old military outpost. Like all other bills of this type, it was referred to the Committee on Internal Improvements.[23] The major piece of legislation for which he should deserve much of the credit during this session was a bill to regulate the militia in the Territory. The Second Seminole War, which was still in progress while the session met, had shown some glaring weaknesses in the structure of Florida's military force. The structure was revamped with specifications on the elections of officers, the appointment of non-commissioned officers by the captains of the unit and the amount of staff the Governor and brigade commanders could be allowed, all subject to approval by the Legislature. More importantly, new requirements were put forth demanding that the drills used by the regular army would be used by the Florida militia. No military man could be arrested by a civil authority while on duty or going to and from duty, "except for treason, felony or breach of the peace." The fines for refusing to do assigned duties were also spelled out and exactly how the courts martial were to function was clearly defined.[24] This reorganization lasted, in its basic form, until the War Between the States and was clearly meant to instill discipline and define the lines of command, which were not well understood during the Seminole War.

On the issues of the banks and the adoption of the St. Joseph's Convention proposed constitution, Francis L. Dancy maintained the line accepted by the Yulee faction. This meant paying at par value the so-called "faith bonds" to maintain the credit of the Territory and also implied the complete acceptance of the idea of statehood with no division of the Territory into two potential states. In almost every instance during the session when roll call votes were recorded, Dancy voted with this group, headed by the "other Senator," James D. Westcott. Dancy was

---

[22]*Journal of the Proceedings of the Legislative Council of the Territory of Florida, 1842* (Tallahassee: Office of *The Floridian*, 1842), 25-26.

[23]*Ibid*, 28.

[24]*Acts and Resolutions of the Legislative Council of the Territory of Florida, 1842* (Tallahassee: C. E. Bartlett, 1842), 25-33. The law has thirty-two sections in it and is one of the longest passed at this session.

squarely opposed to the faction headed by Judge Issac Bronson and Benjamin A. Putnam.[25]

With the Seminole War at an end and his brief legislative career behind him, Francis Dancy returned to his more civilian routine. As Surveyor of Private Claims, he had the responsibility to separate private lands from those of the public. As many of the public land surveys had been done in the areas in which he was to work, this meant going into the field, finding existing marks or lines, advertising in the local papers for claimants or their representatives to meet him in the field, show him their alleged lines and present him with evidence of the claims. Once Dancy had ascertained that the claimant's grant was valid and the lines conformed to those called for in the grant, he would tie the new lines into the lines of public surveys and obliterate the lines of these surveys where they now interfered with the private grant. This task was difficult and politically very dangerous because of the individuals who had either inherited or, more likely, purchased the grants from the original grantee. Dancy also had the dubious chore of working with Surveyor General Valentine Y. Conway, whose instructions to surveyors were among the most bizarre in the history of Florida surveying. Dancy's questions from the field indicate the unclear nature of Conway's instructions. What rules, he asked in one of his first letters to the Surveyor General, governed the surveying of grants which overlapped, as many do in Duval County? If there are no traces of the public surveys in an area, how does one tie the grants into the regular surveys?[26] These were questions which should have been discussed prior to someone leaving for the field. Dancy was in the field for three full surveying seasons, outlasting Conway and working, at the end of his tenure, under the second administration of Robert Butler, under whom the office of Surveyor of Private Land Claims was abolished and its duties absorbed by the office of the Surveyor General.[27]

By the mid 1840s, Dancy had established himself on the shores of the St. Johns River and was beginning to develop as a citrus grower. But, he was constantly involved in the affairs of the day, especially those of a political nature. He was

---

[25] See the *Journal of the Legislative Council*.

[26] *Letters and Reports to Surveyor General*, Volume I: 1825-1847, 501-02. Land Records and Title Section, Division of State Lands, Florida Department of Environmental Protection, Tallahassee, Florida. [Hereafter, Letters and Reports, Volume and page No.]

[27] *Ibid*, 561. This is the last letter sent by Dancy as Surveyor of Private Land Claims. There was nothing personal or political in the abolition of the Dancy's position. It was a very inefficient way to survey the claims and had the potential of putting the two offices at odds.

looked upon as one of the rising leaders of the Democratic Party and was regular communication with his friend, David Levy Yulee. In 1850, he was also considered as a serious contender for the post of Major General of the Florida Militia, opposing Benjamin Hopkins. However, because of Hopkins' reputation and the respect he had earned, especially during the Indian Scare of 1849-50, Dancy had his name taken out of contention.[28] Yet, he did desire some other political posts and was soon asking Yulee for his assistance in obtaining the post of Surveyor General of Florida, for which he was admirably qualified. Fate soon intervened and rewarded him with the political plum of the first appointed State Geologist and Engineer. Yulee advised his friend that this was the position to accept in that it paid a handsome $2,000 per year plus expenses, offered contacts for later employment, a chance to repay certain individuals with the patronage offered by the job, a place to train his eldest son in the engineering profession and, most importantly, an opportunity to keep the family together by having his office in Palatka (read Buena Vista).[29] Dancy took his sage friend's advice.

The position was one needed by Yulee because one of the duties of the State Engineer was to inspect the railroad lines and canals authorized by the Internal Improvement Act of 1855. One of the major requirements of the act was the actual construction of the railroad, in this case Yulee's Florida Railroad, for a specified number of miles, after which the Board of Trustees of the Internal Improvement Fund would allow the sale of the bonds authorized by the act. It was the sale of the bonds that funded the actual construction of the railroad. Dancy was called upon by the Trustees to assure them that the Florida Railroad had met its obligations. The State Engineer did just that and brought down the wrath of Governor Madison Starke Perry, then in the midst of a feud with Yulee of the line of the railroad. Perry believed that Dancy had been less than truthful in his report to the Trustees and that the railroad had not completed as much of the work as Dancy reported. The reason for this alleged falsehood was to allow the railroad to obtain funding to pay for the iron being shipped from New York by the firm of Vose and Livingston.[30] Dancy may have misrepresented the actual con-

---

[28]St. Augustine *Ancient City*, April 13, 1850, 2.

[29]Department of Interior Appointment Papers: Florida 1849-1907. Roll No. 1: Surveyor General [A-D], 1849-1907. Washington: National Archives and Records Service, 1980. M1119. Letter of July 5, 1853. Yulee to Dancy.

[30]*Journal of the Proceedings of House of Representatives of the General Assembly of the State of Florida, 1858* (Tallahassee: Jones & Dyke, 1858), 11-21.

struction, but did not falsely report its grading and alignment. The only fault was the technicality that the iron was not yet upon the crossties, even though it was to be there within days of Dancy's visit to the site.

As this controversy wound down, Dancy sought out Yulee and his other political friends to obtain the office of Surveyor General. The many letters, petitions and other political machinations used by his friends procured the appointment in 1858, replacing John Westcott, the brother of his legislative ally of 1842. With his background as a graduate of West Point (where surveying was taught), a civil engineer, a U. S. Deputy Surveyor and his military experience in the early years, there can be no doubt that Francis L. Dancy was the most qualified man for the position in the State. He took over the position just at the end of the Third Seminole War (1855-58) and had to instruct his surveyors to replace many of the lines obliterated by the Indians. As Surveyor General, he was in the field inspecting his deputies work and in John Dick, he hired one of the more qualified draughtmen in Florida.

His duties required him to manage a substantial office staff, which included the first woman I have found evidence of in State Land records, Martha M. Reid, as his field note clerk.[31] Ms. Reid was his sister-in-law. He also hired his brother R. F. Dancy and his nephew, E. D. Foxhall. Nepotism was rampant in this era and not unusual in any patronage job, like that of the Surveyor Generalship. However, Dancy had to have people he could trust to do accurate, reliable work and people with whom he could work in the cramped quarters allowed in the St. Francis Barracks in St. Augustine, where the offices were located. He also had to appoint the Deputy Surveyors. This led to some difficulties because of the limited labor pool in the state and the fact that politics did play a role in the selection. In one case, Dancy's sometimes blunt nature showed clearly. In writing to Louis Lanier the Surveyor General wrote:

> As to making any addition to your work; I would not feel myself justified in doing so, for the following reasons - From the report of the examination Clerk in this office, on your work, it appears of So loose and bungling character as to afford evidence of gross inattention to your instructions or incompetency. And I would respectfully suggest the

---

[31]"Salary Accounts - 1858-1860 - Surveyor General's Office," 4, 5, 115. Land Records and Title Section, Division of State Lands, Florida Department of Environmental Protection, Tallahassee, Florida.

propriety of your asking to be released from the compliance with the remainder of the contract instead of renewing the bond.[32]

Other surveyors, many of whom he was not personally acquainted with, did remain on the job under his tenure of office because Dancy's examination of their work proved their worth. It was not all patronage and nepotism.

Francis L. Dancy remained on the job until Florida passed the Ordinance of Secession. At that time he closed down the office, sent in his receipts to the General Land Office and even mailed his certified accounts to the United States Treasurer's office asking it to pay the clerks and draughtsmen for their time.[33]

Dancy had no qualms about siding with his adopted State in the impending war. As a recognized leader of the Democratic Party, he backed the Breckinridge-Lane ticket for president and vice president. He participated in the county conventions and served as chairman of the St. Johns County Democratic Convention, which met on September 1, 1860.[34] In November, he also served as part of the committee appointed to draft the business for the public meeting in support of secession.[35] For Dancy, this was not some political lark, but very serious business.

At the outbreak of the War Between the States, Governor Perry appointed Dancy to the dual office of Adjutant and Inspector General of Florida Forces. He remained in that position under Governor John Milton until 1863. He did not make himself popular with certain denizens of St. Augustine, when he noted "there are not twenty men in the city who would volunteer for distant service. They would all volunteer to occupy the Fort and Eat Rations."[36] With supplies becoming a crucial necessity for both the Confederate Army and the general population, Dancy was commissioned as a captain in the Confederate forces and made Commissary Officer in charge of collecting the "tax in kind," which meant chickens, hogs, sugar and agricultural products. While performing these duties, the Union Army led a raid up river from Jacksonville, which, according to his son, James M. Dancy, was designed specifically to capture and ruin Francis Littlebury

---

[32] *Letters of Surveyor General*, Volume 9, 1853-60, 352. Land Records and Title Section, Division of State Lands, Florida Department of Environmental Protection, Tallahassee, Florida.

[33] *Miscellaneous Letters of Surveyor General*, Volume 1, 1860-61, 34-35. Land Records and Title Section, Division of State Lands, Florida Department of Environmental Protection, Tallahassee, Florida.

[34] St. Augustine *Examiner*, September 1, 1860.

[35] St. Augustine *Examiner*, November 17, 1860.

[36] David J. Coles, "Ancient City Defenders: The St. Augustine Blues," *El Escribano*, 1986, 70.

Dancy. Being an ex-West Pointer, a federal official and recognized leader of the Democratic Party, he was a prime target for those wishing to crush the resistance in East Florida.[37] Dancy escaped just in time and was able to watch his family home being occupied by federal troops, who indulged themselves on the already prepared meal.[38] F. L. Dancy, like many other parents in those turbulent times, also had the misfortune of losing one of his sons in battle. Lieutenant Francis R. Dancy was killed at the Battle of Olustee in 1864. No amount of honor or recognition could replace the son lost in the service of the cause.

After the war, Francis L. Dancy returned to his ruined homestead and began anew. His groves produced some of the most unique and flavorful fruit in North America. He was one of the leaders of the Florida Fruit Growers Association and a frequent contributor to magazines and newspapers discussing the process of citrus growing. The poet Sidney Lanier took notice of the Dancy groves in his tour book, *Florida: Its Scenery, Climate and History*, and other writers followed his lead.[39] Dancy became something of a household word when, at his Buena Vista Groves, he developed a new variety of tangerine, appropriately named the "Dancy Tangerine". Noted at the time as a new addition to Florida's impressive citrus industry, one expert proclaimed that "in flavor and external appearance this variety is superior to the original."[40] Dancy's goal was simply put in October 1876, when he wrote, "It is my earnest desire to see the orange industry brought to the highest state of perfection in this state, believing that with proper care and attention, in growing the best varieties and handling the fruit, but a few years will elapse before our oranges will be as common in the European markets as they are at this day in the markets of our own largest cities, and at highly remunerative prices, owing to their great superiority over the oranges raised in most European countries."[41] As a grower and prophet, he was right on the mark.

When Francis Littlebury Dancy died on October 27, 1890, he had left behind a legacy of devotion to duty, his fellow man and, most importantly, his family. His early career building roads and improvements helped to open the migration into the old Southwest. His skill as an engineer made possible the base for the current

---

[37]James M. Dancy, Untitled Memoir Written in June 1933. Typed manuscript available at the Putnam County Archives, Palatka, Florida. See page 5 (Page 3 of the actual memoir).

[38]*Ibid*

[39]Sidney Lanier, *Florida: Its Scenery, Climate and History* (Gainesville: University of Florida Press, 1973, Facsimile of the 1875 edition), 127.

[40]Anonymous, "Nomenclature of the Orange," *Semi-Tropical Magazine*, June 1876, 339.

[41]Francis L. Dancy, "Culture of the Orange," *Semi-Tropical Magazine*, October 1876, 602.

seawall in St. Augustine and the beginning of the preservation of the Castillo San Marcos. Dancy's abilities as a surveyor led to better, more accurate land titles and markings and ended with his appointment as Surveyor General of Florida. His service to his adopted State during the War Between the States cost him a son, major disruption of his family life and the ruination of his groves and plantation. Yet, he persevered through all of the setbacks and reconstructed his life in a way that began a new era for this state. In all phases of his life, he truly was untiring, faithful and efficient.

To the Register of the Land Office at Newnansville E.F.

Under the provisions of the Act of Congress approved on the 4th day of August, A. D. 1842, entitled "An act to provide for the armed occupation and settlement of the unsettled part of the Peninsula of East Florida."

To all whom it may concern:

NOTICE is hereby given that under the provisions of the act of Congress above cited, I, *Sam Reid* do hereby apply to the Register of the proper Land Office for a PERMIT to settle upon *One hundred and sixty* acres of unappropriated public land, lying south of the line dividing townships numbered *nine* and *ten*, south of the base line, and situated as herein described.

I aver that I am *a Married Man Over Eighteen Years of age And able to bear Arms* and that I became a resident of Florida in the month of *April* in the year *Eighteen hundred and twenty five*

I aver that the settlement herein intended is not "within two miles of any permanent military post of the United States, established and garrisoned," at the time of such settlement, and that the same is not known or believed to interfere with any private claim that has been duly filed with any of the Boards of Commissioners, surveyed or unsurveyed, confirmed or unconfirmed.

DESCRIPTION OF THE INTENDED SETTLEMENT.

On the North Side of the Manatee River Twenty Chains from the bank of the River to a Live Oak marked R the Commencing point - thence from Said tree Sixty Chains due North to a Small live oak thence West twenty Six & a half Chains to prairie thence South Sixty Chains & from thence twenty Six & a half Chains East to the beginning Given Under my hand this 1st Jany 1843

Sam Reid

We Certify that the above is a true Copy of the original now on file in this office

Jno W Tapolli Register
Jno Parsons Receiver

CHAPTER 5

COLONEL SAM REID: THE FOUNDING OF THE MANATEE COLONY AND SURVEYING THE MANATEE COUNTRY, 1841-1847

The life of Colonel Samuel Reid is virtually unknown in the Manatee country. Although he was the leader of the colony that led to the eventual settlement of the area, his past has been ignored in favor of those whose families still inhabit the area and who, along with Reid, bravely pioneered the Manatee frontier. This neglect is also the result of the poor image Reid has had as a surveyor of the area. Although vindicated by the Bradens and Robert Gamble at the time, the rumors of poor performance or fraud have persisted. His contributions to the growth of the area, therefore, have been clouded by the mist of the past. It is now time to take a deeper and clearer look into the life of this Manatee pioneer.

Reid entered Florida from Gwinnett County, Georgia, in 1825 and settled in the frontier area near Tallahassee.[1] In 1833, he purchased forty acres southeast of Tallahassee in Township 1 South, Range 2 East.[2] It does not appear that he was too interested in farming as a lifetime work, and in 1837, he purchased Lot 170, original plat of Tallahassee, on the corner of Jefferson and Monroe Streets, two blocks north of the capitol.[3] On June 7, 1838, Reid entered into a partnership with

---

[1] Armed Occupation Permit No. 316. Newnansville Land Office. Land Records and Title Section, Division of State Lands, Department of Environmental Protection, Tallahassee, Florida. A copy of this permit also exist in the National Archives, Suitland Research Center, Suitland, Maryland, in Record Group 49.

[2] Deed Book C, 374. Leon County Property Records, Microfilm No. 10. Leon County Clerk of the Circuit Court. Microfilm in the Florida Department of State, Division of Archives and Records Management, Tallahassee, Florida. (The property was purchased from John Methina.) Hereafter, Deed Book Letter and page number.

[3] Deed Book E, 706.

James B. Gamble and began to sell "a general assortment of goods." The new firm operated under the name of Gamble & Reid.[4] The 1839 tax rolls showed that the firm had one slave working in the store and an inventory of $18,000, quite large for the era. It also showed that the town lots on which the store sat were valued at $3000, for which $32.50 was paid in Territorial taxes and $66.00 in County taxes.[5] The firm lasted only a year and a half and was dissolved by mutual consent on January 1, 1840. The firm continued to do business under the name of James B. Gamble, until he was joined by J. Gratton. Gamble, when the name was changed to James B. Gamble & Co.[6] The same issue of the Tallahassee *Floridian* that announced the formation of the latter firm advertised that Sam Reid was going into the "Storage and Commission" business at Port Leon, the new terminus of the Tallahassee Rail Road.[7] In fact, Sam Reid was the first purchaser of lots in the new city.[8] By mid-1840 he was doing an active business at Port Leon and owned "extensive Warehouses and Wharf" in that town. These facilities he leased to a J. Vail in early 1841.[9]

In January 1835, he must have been well enough settled to take as his bride, Carolina J. Alston, on January 7 of that year.[10] The Alston family was one of the more influential families in Middle Florida at this period and the marriage brought Reid into a wider circle of powerful people. His brothers-in-law included Dr. John Bacon and David S. Walker, later governor of Florida and one of Reid's closest confidants.[11] The Alston's were the family involved in the famous duel and

---

[4]Tallahassee *Floridian*, February 9, 1839. The advertisement was dated June 7, 1838, announcing the new firm.

[5]"Tax Rolls Leon County 1829-1855," (Incomplete file). Microfilmed by the Geneological Society, Salt Lake City, Utah. 1956. Copy at the State Library of Florida, Florida Room (Dodd Room), Florida Department of State, Tallahassee, Florida. Page number is unreadable.

[6]Tallahassee *Floridian*, January 11, 1840 and January 15, 1840.

[7]*Ibid*

[8]Deed Book E, 836. This shows Reid purchased Town Lots 6 and 7, Block 1, complete with water privileges. He purchased these lots for $5 each. Also see, Elizabeth Smith's special edition of the *Magnolia Monthly* entitled "A Tale of Three Tombstones." (Crawfordville: Magnolia Monthly Press, 1968), 27.

[9]*The Florida Sentinel*, May 28, 1841. The advertisement stating this information is dated April 8, 1841, and was run in successive editions of the paper.

[10]Tallahassee *Floridian*, January 10, 1835. Found referenced in "Leon County Marriages", Florida Room, State Library of Florida, Tallahassee, Florida. This is a loose index taken from contemporary newspapers and bound for reference work. No date of publication.

[11]Walker married Philoclea Alston on May 24, 1842, while Bacon married the third sister, Clementina, on May 24, 1837. "Leon County Marriages." Florida Room, State Library of Florida, Tallahassee, Florida. Later letters from Reid to the Surveyor General of Florida often asked that gentleman to forward his personal letters to his wife through Walker.

murder involving General Leigh Read. Indeed, Sam Reid attended a special meeting in honor of the late Augustus Alston, killed by Leigh Read, in late December 1839. Also attending the memorial meeting were James B. Gamble, Robert Gamble, D. S. Walker, R. B. Ker and Arthur M. Randolph, son-in-law of former governor William Pope Duval.[12]

The 1840 census showed that Reid had three children under the ages of five, two daughters and one son, living with him and, in addition to his wife, one other adult between the ages of thirty and forty. The same document also recorded that he owned fifteen slaves, eleven males and four females. The household, therefore, totaled twenty-one persons according to these figures.[13] To support such a large number of individuals means that Reid was somewhat successful in his business operations.

Reid's other interests are very noteworthy. In 1834, for example, as a stockholder in the Tallahassee Rail Road, he signed a petition on behalf of the line asking for federal lands. The president of the railroad was Governor Richard Keith Call.[14] He was one of the signers of a number of resolutions sent to Washington in 1838, along with William P. Duval, A. M. Randolph and R. B. Ker.[15] Most importantly for the future of the Manatee area, on February 24, 1840, he signed a petition supporting the concept of law establishing military colonization of the Florida frontier, which had been proposed by Governor Call. Also signing this lengty petition were six of the Gamble family, William H. Wyatt, John Addison of Gadsden County, and two members of the Grisset[h] family, also from Gadsden County.[16]

However, this picture of success and political activity must be tempered by the fact that, by 1843, his operations in Port Leon had not been prospering for reasons unknown. In that year, Sam Reid was forced to convey title to William Bailey, of Jefferson County, to his property in Port Leon, in addition to five slaves—May and her four children. The transfer of this property was through a default on pay-

---

[12]Tallahassee *Floridian*, December 21, 1839.

[13]1840 Census for Leon County, Florida, within the division allotted to George E. Dennis. 65. A microfilm copy of this record is located in the Florida Department of State, Division of Archives and Records Management.

[14]Clarence E. Carter, Editor, *The Territorial Papers of the United States: The Territory of Florida*, Volume XXV, 1834-1839 (Washington: Government Printing Office, 1960), 77. Hereafter, *Territorial Papers*, volume and page.

[15]*Territorial Papers*, XXV, 464-66.

[16]*Territorial Papers*, XXVI, 81-88.

ments on two notes totaling seven thousand dollars and backed by George K. Walker, brother of David S. Walker. If Walker met certain conditions, however, the transfer was to become null and void. As holder of Reid's notes to Bailey, Walker stood to gain the Port Leon property if he would make the payments.[17] This venture probably went sour when Port Leon was destroyed by a hurricane later in 1843.

About April 8, 1841, Reid leased his business interests in the Port Leon warehouses and wharf to J. Vail and headed south to Tampa Bay, having accepted the position as Deputy Collector of Customs for the Port of St. Marks. On July 1, 1841, he wrote to R. W. Alston, his brother-in-law, about the need to send a revenue cutter to the coast of Florida to prevent the Spanish fishing camps from selling arms and supplies to the Indians. Reid's letter was passed on to former governor William P. Duval who sent it on the Secretary of Treasury. Duval, an old acquaintance of Reid's, noted, "Mr. Reid is an intelligent man, of high character, and a most vigilant [sic] officer." The fact that Reid's views coincided with those of the ex-governor was an important factor in the transmission of this letter.[18] The letter demonstrated Reid's close contacts with the politically powerful families and their recognition of his worth.

Reid's duties as Deputy Collector of Customs required him to become acquainted with the area, which stretched from Charlotte Harbor northward. As such, he most likely scouted out the area of the Manatee River, at the time a virtually unsettled wilderness. His post and ownership of warehouses at Port Leon brought him into contact with the staff of General William Worth, then commanding the United States army in Florida. Through Worth's liaison, Lieutenant M. Patrick, Reid was recruited to lead a colonization effort on the Manatee River. The effort began on April 16, 1842, when the little band of colonists, headed by "Colonel" Samuel Reid landed at Manatee. According to General Worth, the colony "is composed entirely of persons from Middle Florida. The land is of Superior quality & from the character of the Gentlemen concerned, there is certainty of success." The General also noted that the colonists had been issued arms. (200 "Ball Buckshot & Cartridges," 10 muskets and 20 musket flints, along with

---

[17]Record Book of Hillsborough County, Territory of Florida, Vol. III: 1838-1846. Copy prepared by Historical Records Survey, Works Progress Administration, Jacksonville, Florida. 1938. 376-78.
[18]*Territorial Papers*, XXVI, 363-64.

tents.)[19] The new settlement totaled fifteen white males, ten black males, two black females over fourteen and four black children, for a grand total of thirty-one individuals.[20] Reid's scouting of the area, possibly in company with Josiah Gates and others, proved to be important in establishing the colony on the river and showed his keen sense of judgment.

The army units posted at Tampa Bay's Fort Brooke were to act as guardians for the young colony. On May 18, 1842, Assistant Adjutant General T. Cooper wrote to Major T. Staniford, then commanding at Fort Brooke, "The Colonel Commanding desires you will consider the party under Col. Reid at Manatee river, in all respects on the footing with others the most favored, & to afford them every facility & encouragement. He desires you will furnish them Arms &c. as a loan, to be accounted for by Lt. Patrick, to whom report will be made in the case. The tents loaned the party are to be retained by them until they can conveniently house themselves, so as not to interfere with the planting of crops."[21] This stewardship was to prove mutually useful to the settlers and the army in the coming months and years.

Reid's leadership role involved him with the military as the eyes and ears of the colony and adjacent frontier. He was frequently required to quell false rumors and assure the colony, and others, that the Indians posed no threat to their existence. In September 1842, a truce with the Indians was concluded and General Worth wanted the frontier settlements notified that Indians would be moving through their locales. Reid was informed by a letter that Indians would be moved through the Manatee area, on their way to embarkation at Tampa. Lieutenant P. A. Barbour wrote, on behalf of Worth, to the Commander at Fort Brooke, Captain William Seawell, "Colonel Reid has been written to today and advised of the intentions of his southern neighbours to visit Tampa. His settlement need not be

---

[19]Report of June 13, 1842. Worth to General R. Jones, Adjutant General. Letters Received by the Office of the Adjutant General, 1822-1860 (Main Series). Record Group 94. Roll 262. W 217-405. National Archives Microcopy 567. 1842. The colonists of this colony are listed as follows: 1. Col. S. Reid; 2. J. Gates; 3. Mr. Ledwith; 4. Mr. Price; 5. F. Follansher; 6. Mr. Retterline; 7. Mr. McDonald; 8. Mr. Craig; 9. Wm. H. Wyatt; 10. John Bowers; 11. Dan. Buchanan; 12. John Addison; 13. John Griseth.

[20]*Ibid*

[21]Letter of May 18, 1842. Cooper to Staniford. Letters Received by the Office of the Adjutant General, 1822-1860 (Main Series). Record Group 94. Roll 262. W 217-405. National Archives Microcopy 567. 1842.

visited from your post."[22] As the leader of the colony, Reid was informed of most of the moves of the military and their possible impact on his settlement.

Like many, but not all, of the settlers, Reid received an Armed Occupation Permit (No. 316, Newnansville Land Office) to occupy the land he had already begun to clear and plant. The date of the founding of the colony—April 16, 1842—precedes by many months the passage of this landmark piece of legislation. Because of this, it is likely that some of the settlers who accompanied Reid left prior to the receiving or requesting such a permit. Some may have died, too, however, there has yet to be uncovered any record of this. It may also be that the success of the colony did encourage many of the Armed Occupation Act settlers to try their luck on the Manatee River. These speculations, however, lack documentary proof at the present time. The only sure thing we can note is that many of the settlers who came to Manatee with Reid, did receive their permits and patents under this act.

Reid continued in the leadership of the colony when an incident, typical of the frontier, occurred. In mid-1844 rumors were flying that another Indian war was looming and that Indians had been spotted near the settlements. Allegedly one of the survey party of Henry Washington's had reported seeing many Indian signs in the vicinity of the Manatee River, where that crew was working. On August 5, 1844, General Worth wrote the following to Captain Montgomery, commanding at Tampa, "Sir: I desire you to cause the enclosed communication to be forwarded with the least avoidable delay to Colonel Reid, by the Star, if at Tampa, and not urgently employed, instruct the messenger to await for Colonel Reid's answers if he can be found at his residence...Seal the letter to Colonel Reid before forwarding." The same letter noted that the affidavits requested by General Worth to the falseness of the Indian scare, must be done quietly and without arousing any undue suspicion on the part of the settlers.[23] The rumor started by a "drunken scoundrel" of Henry Washington's survey party, was quickly squelched and ended. Worth suspected that the entire story was a pre-planned episode meant to arouse the settlers and the government to attack the Indians and drive them, finally, from Florida.[24] The point of interest in this correspondence is the contin-

---

[22]Letter of September 14, 1842. Barbour to Seawell. Letters Received by the Office of the Adjutant General, 1822-1860. (Main Series) Record Group 94. Roll 260. U-W 46. National Archives Microcopy 567. 1842.

[23]Letter of August 5, 1844. Worth to Montgomery. William Worth Belknap Papers, Box 1, Princeton University, Princeton, New Jersey.

[24]Letter of July 7, 1845. Worth to Lt. Colonel Belknap. William Worth Belknap Papers, Princeton University, Princeton, New Jersey.

ued importance of Sam Reid as the recognized leader of the colony, even though the Bradens, Gamble, Gates and others were already on the scene.

Shortly after the founding of the colony, it was recognized by the government that the Manatee area would soon have to be surveyed to assure proper title to the lands of the colonists. Sam Reid requested the appointment as U. S. Deputy Surveyor and called on friends to support his application. Richard Keith Call responded to the this request and wrote to the Surveyor General, "my friend Col. Saml Reid, who is anxious to obtain a contract to survey a portion of the public lands in Florida. He is in every respect worthy and well qualified."[25] With the help of such friends, Sam Reid became the U. S. Deputy Surveyor when he signed his first contract on November 21, 1843. For his new duty, he received $3.75 per mile of survey line, $.25 lower than the average for the day.[26]

Reid immediately requested an Army escort into the area of his survey because of the presumed Indian threat. The Army, however, refused to allow this as the escort that accompanied Henry Washington's crew proved to be counterproductive, scarring Indians on the way to surrender back into the Big Cypress and out of the reach of the troops.[27] Reid's first survey was also delayed because of "incessant rains and high waters."[28] Reid described the conditions at that time, "I repaired early in January to the field and have been constantly and labourously engaged ever since at it, but from the quantity of Rain which fell about that time, I find it impossible to commence at the Southern boundary of my district as the whole Country was overflowed. Before I ceased my party were frequently on the point of Starvation for water."[29] The situation on the frontier was wet and threatening for the new surveyor, but, undaunted by these temporary setbacks, he pushed on to complete his contract. What is important to note here is the fact that Reid did *not* head to the southern end of the contract area, but finished the number of required miles in the northern end of the district. Before the year was out, he was to be accused of running false lines in the area he clearly did not survey.

Surveyors had numerous obstacles to completing contracts and making ends

---

[25] Letters of Application. Voume 2, 1825-1847. Land Records and Title Section, Florida Department of Environmental Protection. Tallahassee, Florida. Hereafter DEP.

[26] Contract File of Samuel Reid. Drawer: U. S. Deputy Surveyors O-Z, File: U. S. Deputy Surveyor Samuel Reid. Land Records and Title Section. DEP.

[27] *Letters and Reports to Surveyor General*, Volume 1, 1825-46, 265-69. Land Records and Title Section. DEP. Hereafter, *Letters and Reports*, volume number and page number.

[28] *Territorial Papers*, XXVI, 654.

[29] *Letters and Reports*, Volume 1, 277.

meet during the surveying season. Two of these have been noted above, however, added to them were the problems of recruiting a capable crew that could be counted upon to endure the hardships and competently fulfill the needs of the surveyor. The surveyor also had to purchase instruments, supplies, wagons, mules/horses, field books, and other supplies before embarking on the venture. This meant that most of the deputy surveyors had to have some up front money before taking on any contract. Additionally, each surveyor had to be bonded and have someone willing to underwrite the enterprise. This required having contacts willing and able to put up the bond money, or its equivalent, before the survey could start. The four to six month surveying season, done in "dry" season only, and the constant possibility of sickness, death, insects, injury, or other mishaps added to these difficulties. All of these things made surveying a very speculative business and some surveyors did lose their investments by not completing the contracts on time, if at all.

On August 22, 1843, Reid wrote that he had arrived home and reported that not much had been done and that he had been becalmed four days on a small sloop. He then noted that he was beginning the chaining of sections in Township 34 South, Range 17 East, but the rainy season prevented much from being accomplished. He did tell Surveyor General Valentine Conway that he would send his notes via Colonel Braden as soon as that gentleman returned to Tallahassee. Indicative of his interest in the development of the colony, Reid also included samples of the first tobacco crop raised by the colonists.[30] By November 15th, he had completed the survey and had sent in the notes for approval.[31]

At this time, Reid notified the Surveyor General that there was a potential conflict in the Armed Occupation claims of Mr. Ledsworth and Mr. Price, both of whom were absent from the area at the time of the survey. Josiah Gates, the brother-in-law of Price, sent a note to Reid identifying his relative's claim and that of Mr. Ledsworth. Reid asked for the discretion to adjust the claims so that they did not fall in the same quarter section of land, but specifically asked for specific instructions on how to make the alteration so as not to take any of the improvements from either party. For the information of the Surveyor General, Reid also notified him of new settlers coming into the area, after the period for the Armed

---

[30]*Letters and Reports*, Volume 1, 280.
[31]Letter of November 15, 1843. Reid to Conway. Land Office Notices, Refusals, Acceptances and Sundry Letters (file) Armed Occupation Permits M-Z (drawer). Land Records and Title Section. DEP.

Occupation law, most of whom, he noted, settled in the pine lands.[32]

In late October of 1843, the Surveyor General had written Reid informing him of the charges of false surveys being levied against him. However, because he was in the woods surveying, he did not receive the information until January 8, 1844, a delay of over two months. Reid immediately informed Conway:

> I regret exceedingly that I had not recd your letter earlier, as I could long since have satisfied you that the charges against me are entirely false. You will see by my returns that no Surveys now are reported to have been made at or near Charlotte Harbour. I saw Colo. Washington at Tampa last winter, just as I commenced work, who informed me that his line terminated in Town. No. 40 and the no. of miles, but I have forgotten the distance, but I am under the impression that the line running West between Township 38 & 39 is not less than ten miles of any portion of Charlotte Harbour...But Sir the whole statement of those Alabama gentlemen is false, I assert it and believe that the field notes of Col Washington will Sustain me, that after going South 2 1/2 miles in Town. 37 Range 22 that there is not five acres of land, that is not in ordinary wet season covered with water and that except some Cypress Swamp on Peas Creek, there is not one acre of hammock land in or about Charlotte Harbour...He [Washington] told me of this when I saw him, and he advised me to throw up mounds, and to prepare myself with a spade to do so. I did as he directed....I have Sent up requesting Colo Braden and indeed the whole neighbourhood to come around and examine the work."[33]

Of course, Joseph Braden did come to the aid of his friend and neighbor, writing to Conway on January 17, 1844, "I have seen a letter from you to Col Reid that he is charged by some Gentlemen from Alabama with making 'sham' surveys. I have been a resident on the River for a period commencing within a few days of his surveys until the present time, & am satisfied that the persons who made these charges, have never been on the River, & not have recd the information <u>here</u>, that they were 'sham' surveys. Such is not the opinion of those living on the River, & who have had many opportunities of ascertaining whether the surveys were worth making or not. I have no hesitation in saying that the charges are malicious & groundless."[34]

---

[32]*Ibid*
[33]*Letters and Reports*, Volume 1, 281-82.
[34]*Letters and Reports*, Volume 1, 285. See also page 293 of same volume.

The alleged group of Alabama gentlemen was a hoax hatched by a disappointed surveyor, who had been seeking employment with the Surveyor General and with Reid, but was rejected because of his lack of proper character and behavior. The letter accusing Reid is a good example of creative thinking. As Reid had written, the description offered by the "Alabama Gentlemen" was pure fiction. The land claimed to have been seen and marked by these people is described as "varied and picturesque scenery of rich hammocks, prairie, and pine lands interspersed with ponds and bayous [sic], which enhanced it in our estimation as a first rate grazing range for cattle."[35] Anyone, to this day, familiar with the area of Charlotte Harbor can see the falseness of this description. Yet, Reid was called upon to defend his surveys and his reputation against the slanderous attack.

In the October 26, 1843, letter to Reid, the Surveyor General advised him on how to handle the charges and prove the validity of his work. He recommended that he take some of the local, reputable people out to his work, let them examine the marks and lines and then furnish sworn affidavits to what they had seen. Conway concluded, "This course may have the tendancy to disabuse the minds of all interested and supersede for the present the necessity of commissioning another Deputy to go in and examine your work."[36] Reid followed this good advice and got the cooperation of Judge Josiah Gates and Hector W. Braden who swore that "We made a particular examination of these several lines, amounting to more than nine miles, including ten corner posts, and more than fifty bearing trees; these we found well marked and easy to delineate. We find no difficulty in following any of these lines or ascertaining with facility the Townships Ranges and Sections on the entire route."[37] These were strong witnesses for the surveyor and their influence proved important in finally ending the speculation regarding the correctness of his work.

The man behind the accusations was Robert B. Ker, a man known to Reid and many of the early settlers of the Manatee River region. Ker had been active in many of the social and political events in Tallahassee and served as Deputy Surveyor on many occasions, including the final survey of the boundary of the Forbes Purchase. However, the job of Deputy Surveyor was a political appointment, in most instances, and Ker was not in with the group around Valentine Conway. Conway, after some initial hesitation, saw the evidence Reid had referred to and

---

[35]*Letters of Commissioner*, Volume 3, 1840-43, 600-702. Land Records and Title Section. DEP.
[36]*Letters of Surveyor General*, Volume 4,1842-44,102-03. Land Records and Title Section. DEP.
[37]*Letters and Reports*,Volume 1, 297.

the affidavits of Braden, Gates and others and was convinced that Reid's work was legitimate. It was soon suspected that the entire episode was being staged by Ker in revenge for being refused employment, especially since none of the seven signees on the petition from the "Alabama Gentlemen" were known to anyone, including those on the Manatee River supposedly interviewed by these men.[38]

In what appears to be a final desperate act by Ker, he wrote to David Levy Yulee, Florida Delegate to Congress, stating the same case alleged by the "Alabama Gentlemen." Ker even stated that he had confidence in one Charles D. Chesterfield of this group and believed Chesterfield had a basic knowledge of surveying and was able to correctly judge the work done by Reid. He concluded his tirade by, again, stating the impossibility of running 800 miles of lines in four months, which many surveyors claimed to have done in Florida.[39]

When Conway was sent a copy of Ker's letter to Yulee, he was quick to respond:

> The Author of this communication applied to me for a contract in the fall of 1842. On instituting an enquiry into his character & standing I soon learned enough to prevent me from complying with his wishes. Indeed, on one of those occasions he presented himself before me in a high state of intoxication and subsequently I have frequently observed him in a similar situation. Chagrined and disappointed in his application he has sought revenge by attempting to cast odium upon the work executed by my Deputies in the field…Now Sir, after the most diligent enquiry I cannot ascertain the actual whereabouts or identity of an individual member of the company of disappointed & disaffected Explorers of Hammock Land & Marks & believe me when I assure you that it is and has long been my firm conviction that R. B. Ker, himself is the getter up & concoctor of this whole scheme of defamation & falsehood with design to injure Col. Reid and bring into disrepute the surveys generally.[40]

Reid too, found out the author of this cruel hoax and challenged Ker to come to Manatee and show where these "sham" surveys were. Reid went so far as to offer to pay Ker's travel expenses.[41] Ker did not take up this challenge.

---

[38] *Letters of Surveyor General*, Volume 4, 73-75.
[39] *Letters of Commissioner*, Volume 4, 17-19.
[40] *Letters of Surveyor General*, Volume 4, 113-16.
[41] *Letters and Reports*, Volume 1, 303-04.

Commissioner of the General Land Office, Thomas H. Blake, effectively ended the controversy after obtaining other evidence on the character of Robert B. Ker. He had the correspondence of Reid and Ker before him when he made the decision and also had affidavits from Braden, Gates and others. Blake complimented Conway on the manner in which he had handled this small crisis and maintained the public confidence in the surveys. He informed Conway that he was totally satisfied with the correctness of Reid's surveys and had acted upon his accounts. Payment for which was already on the way to Reid.[42]

By mid-1844, Reid was again in the field trying to survey some of the coast near Teira Cia Bay. This was difficult surveying because the land was so broken, judgment had to be used in determining what islands had enough land to pay for the cost of surveying and the nature of the tides complicated these judgments. He also noted that a previous surveyor in the area of Township 27 South, Range 18 East had not followed instructions correctly and had thrown excess lands onto the southeast or southwest. As the range lines had been run from north to south and he had started from the southeast corner and run north in sectioning, the two lines did not match, which he correctly noted, made his surveys look bad.[43] Reid was well aware that his surveys were under tight scrutiny and made every effort to run his lines correctly. Yet, diligent as he was, the taint of the Ker investigation, the lack of remaining monumentation within a decade and the fact that his contacts were politically powerful have clouded the judgment of some as to the correct nature of the majority of his work. The majority of today's surveyors believe that Reid's work was relatively accurate, however, some still have doubts about his ability after all these years.[44]

Reid's family moved to the Manatee area, probably in 1844, from the Tallahassee area, and remained until shortly after he died, in April 1847. Evidence of this occupation of the land is found in the records of Leon County, where it is recorded that Carolina S. Reid "of Hillsborough Co." bought the crops of Robert Alston, her brother, for a tidy sum of $15,000. The crop, the majority of which was cot-

---

[42]*Territorial Papers*, XXVI, 891-92.

[43]*Letters and Reports*, Volume 1, 307-08. Letter of April 15, 1844.

[44]Discussions in my seminars on the History of Surveys and Surveying, conducted for the Florida Society of Professional Land Surveyors, have given me valuable insight into today's opinions concerning Reid's work. Two seminars in Tampa and one in Sarasota have given me the opportunity to talk to the majority of surveyors who have attempted to follow his field notes. My colleagues in the Bureau of Survey and Mapping also have found Reid's work, by and large, fairly accurate. There are, however, one or two strong dissenters from this opinion.

ton, was to be sold to pay off this debt, and any shortages would be made up from the next year's crop.[45] However, Carolina Reid left the Manatee area after Sam's demise for we find her, again, on the records of Leon County purchasing land and crops near Lake Miccosukee.[46] It would appear from this evidence, none of Sam Reid's family remained in southern Florida after his death.

The next to last letter we have of Sam Reid, was penned on November 10, 1846. In this letter he clearly knows that he is dying. At the same time, he also passes the torch into the hands of the capable John Jackson:

> This will be handed to you by Mr. John Jackson, who visits St. Augustine for the purpose of making my returns, my continued illness making it impossible for me to do it in person. Mr. Jackson has been with me through the survey, and can give you any and all information which you may require touching the survey. You will find Mr. Jackson a scientific, intelligent and honorable man, and every way worthy of any confidence you may place in him. I fear that I have run my last line, as my protracted illness gives me no room for hope for a speedy recovery, if I recover at all, and would therefore recommend Mr. Jackson to your favorable notice, as an accomplished surveyor.[47]

Reid's ability to pick capable people to do certain jobs or join in colonization efforts proved to be uncanny. Most of the early settlers, as we have seen, were brought together by Sam Reid and remained to found the prosperous Manatee colony. These hardy men and women were the backbone of the colony and the pioneers of the area. Unfortunately, the name of the man who brought them here has remained forgotten until now. His choice of John Jackson, one of Florida's most accurate and dedicated surveyors, was a final note to his ability to see the true and necessary character of those he associated with on the frontier of southern Florida. Without his abilities, persuasiveness and tenacity, the Manatee colony may not have been as successful as it proved to be. Thus, we should now add the name of Sam Reid, the true founder, to the list of valiant pioneers who established one of Florida's premier settlements.

---

[45]Deed Book I-J. 259. (Dated May of 1847.)
[46]Deed Book I-J. 381. (Dated April 28, 1848.)
[47]*Letters and Reports*, Volume 1, 313.

CHAPTER 6

A SURVEYOR'S LIFE: JOHN JACKSON IN SOUTH FLORIDA

John Jackson, pioneer surveyor, general store owner, mill owner and civic leader has had comparatively little attention paid to his remarkable career.

Few pioneers can claim that they laid out the boundaries of a major metropolis, avoided Indian attacks in the wilds of the Everglades, owned and operated a general store and mill and took part in many civic affairs culminating in a term as mayor of the city whose boundary he helped to establish. Many of the early citizens of Tampa owed their property descriptions to this Irish immigrant and engineer. In a long and distinguished career as a professional surveyor and civil engineer, he often defended his work and advised others until well into advanced age and long after he had retired as an active surveyor. Such a memorable person deserves more recognition in the eyes of history.

John Jackson was born in 1812 at Ballybag, County Monaghan, Ireland, the son of Hugh and Ann (Corocran) Jackson. His early education appears, from the limited records, to have been obtained in the local schools of County Monaghan and rounded off, in the fashion of the day, in some form of apprenticeship to a local engineer.[1] By 1841, the economy of Ireland was fairly depressed and the beginnings of famine already appearing, John Jackson and his younger brother, Thomas, immigrated to the United States. The Ireland he left was heavily populated and arable land was too expensive for anyone but English lords. The population was so high that Benjamin Disraeli declared that the country was more densely populated, on its usable land, than China. Even some of the census takers of Ireland thought that the numbers (over eight million) used by the official record were far too low. Unemployment was widespread and the specter of people living

---

[1] *Florida Genealogical Journal*, 18 (1982), 43.

John Jackson
(Florida State Photographic Archives)

Catherine Maher Jackson
(Florida State Photographic Archives)

in caves and sod-huts surely influenced many to leave the Emerald Isle.[2] The Jackson brothers soon found themselves in the bustling port city of New Orleans and John made his living as the assistant city engineer. It was during this period that John Jackson met Simon Turman. Turman, who was then heading a group of colonists migrating to Florida, persuaded Jackson to join the group and avail himself of the opportunity to receive 160 acres of free land, which were offered to settlers under the Armed Occupation Act of 1842. To the young Irish engineer, 160 acres of free land seemed too good to be true, given his background. Jackson readily accepted the offer and came to Tampa Bay in early 1843.[3]

The Armed Occupation Act of 1842 authorized an individual to stake a claim on government land if they agreed to clear at least five acres of land, build a house suitable for human beings and to bear arms against any Indian aggression. The purpose was to create a line of armed settlements on the frontier between the Indians remaining in Florida and the main white settlements and towns.[4] Jackson, Turman and many others eagerly sought out these lands on the frontiers of modern Hillsborough and Manatee Counties. Jackson's first permit, No. 917 Newnansville, requested the Southeast corner of Section 2, Township 34 South, Range 18 East. However, there was a conflict with a claim already filed by one William Mitchel and Jackson withdrew his request for this land and settled for the Southwest corner of Section 13, Township 34 South, Range 17 East.[5] This land was near Turman's on the Manatee River, but, by 1845, both men had moved to Tampa and had begun relatively successful careers in their new homes.[6]

While living on the Manatee River, Jackson met and was befriended by United State Deputy Surveyor Sam Reid. From this friendship bloomed a new and lucrative career for John Jackson, that of a U. S. Deputy Surveyor. It was Reid who introduced Jackson, through letters, to Surveyor General Robert Butler, a former acquaintance of Reid's from his days in Leon County and as his superior in the surveying business. By late 1846, Butler had contacted Jackson to survey on his

---

[2]Cecil Woodham-Smith, *The Great Hunger: Ireland 1845-1849* (New York: Harper & Row, 1962), 23-31.

[3]*Florida Genealogical Journal* and Karl H. Grismer, *Tampa: A History of the City of Tampa and the Tampa Bay Region of Florida* (St. Petersburg: St. Petersburg Printing Co., 1950), 105-06.

[4]James Covington. "The Armed Occupation Act of 1842," *Florida Historical Quarterly*, 40 (1961). This article contains the best account of the provisions of the act.

[5]Armed Occupation Act Permit # 917. Newnansville Office. Undated letter found in: Drawer N-Z, Armed Occupation, File "Land Office Notices, Refusals, Acceptances, and Sundry Letters," Land Records and Title Section, Florida Department of Natural Resources, Tallahassee, Florida.

[6]Grismer, *Tampa*,106.

own as an official U. S. Deputy Surveyor.

Jackson's friendship with Reid, the subject of some sharp criticism and charges of fraudulent surveying in the Manatee region, lasted until the latter's death in 1947. But, another relationship grew directly out of the circumstances into which Reid had enticed Jackson. Because the Surveyor General's office was located in St. Augustine, Jackson, to file his Field Notes and returns, had to travel to the Ancient City. It was while on such a duty that Jackson met his bride to be, Ellen Maher, the daughter of Robert and Catherine Maher of County Tipperary, Ireland. The arrangements were quickly made between the two and they were married on July 22, 1847. This marriage lasted until John Jackson's death in 1887 and she continued to live in her Tampa home until her death in January 1906.[7] Thus, by conducting the business of surveying for the government, Jackson's friend had introduced him, albeit indirectly, to the person destined to be his closest friend and companion, Ellen Maher Jackson.

Jackson's first major job as a surveyor in the new area was to lay out the town of Tampa. The town had allegedly been platted by Judge Augustus Steele in 1838, but it appears that this work was not actually completed except for Tampa and Water Streets. Jackson was given the job of completing the survey and extending it into the new areas of settlement. In the process, he named many of the early streets of Tampa, mostly after presidents and military leaders. The survey took just a little over two months to complete and the town plat was recorded officially on January 9, 1847.[8]

Shortly after John Jackson finished his work in Tampa, Surveyor General Butler sent him to survey the private grant given to the late Dr. Henry Perrine on the Southeast coast, near the city that presently bears his name. This survey introduced Jackson to a whole new type of terrain that was not very appealing to the surveyor. Jackson wrote to Colonel Butler on June 12, 1847, "This is a very Rocky country we can wear out 2 pairs of Shoes (each of us) every week notwithstanding all this there are some tracts of very fine rocky firm land."[9] He had trou-

---

[7]*Florida Genealogical Journal*, 32.

[8]Grismer, *Tampa*, 106.

[9]Letter of April 5, 1847. Jackson to Butler. *Letters and Reports to Surveyor General*, Volume 1, 1825-1847, 819-20. Land Records and Title Section, Florida Department of Natural Resouces, Tallahassee, Florida. The letters to the Surveyor General are bound into three volumes and are a fruitful source of primary information regarding the conditions of surveying the Florida Frontier. For the sake of brevity, they will be referred to only as *Letters and Reports*, volume number and page number, if given.

ble finishing the contract on time because, "the Country is so rough and (in the latter part of the time) so wet; that I could not get done sooner."[10] While waiting for representatives of the heirs of Dr. Perrine to show him the approximate location they desired, Jackson surveyed some additional mainland property and marked off the three acres that were to become the rebuilt Key Biscayne lighthouse.[11] Thus, John Jackson not only laid out the important Perrine Grant but also surveyed the sight for one of Southeastern Florida's major historical sites, the Key Biscayne/Cape Florida lighthouse.

While on this survey, Jackson informed Butler of some technical difficulties he had with the variations used by the previous surveyor in the area, George Mackay. As he wrote his returns, he notified the Colonel that he would be in St. Augustine within three to four weeks to file his report and Field Notes. These two small isolated notices indicate the larger problems faced by the professional land surveyor of early Florida. In the first place, there was the technical competency needed to follow the directions of the Surveyor General and apply them in the field. Secondly, the surveyor, at his own expense, had to procure a survey team, outfit them, find transportation, file his bond, get provisions for the crew and get into the field and begin work. Upon finishing the field work, the surveyor would then have to pay the crew and other outstanding expenses, correct his field notes and sketches, get his accounts squared away (miles surveyed and meanders run) and then travel to St. Augustine to file the finished product with the Surveyor General. If the Surveyor General found any errors or miscalculations, etc, he would return the work for personal corrections by the surveyor. Should the surveyor be fortunate enough to pass muster with the Surveyor General, he then ran the gauntlet of the General Land Office in Washington, which could accept or reject the work on any technicality. The Comptroller then reviewed his contract, bond and expenditures to make sure they met the standards of the day. Not until all of this was completed was the surveyor compensated for his work by a draft drawn on a regional bank. As Florida's banking system was nearly nonexistent in the 1840s, this meant the drafts were drawn on a regional bank, either a Savannah or Mobile based bank. The entire process could take as much as a year to complete and sometimes took even longer. By implication, the surveyors had to be either men of some wealth or someone who had a good standing in the community who would be backed by citizens with the means to support the survey. The system was open to a variety of

---

[10]Letter of June 12, 1847. Jackson to Butler. *Ibid*, 824.
[11]Letter of April 5, 1847. Jackson to Butler. *Letters and Reports*, 1, 815-16.

pressures which, to use the modern term, could lead to "insider trading."[12]

In late 1848, John Jackson was again called upon to survey the immediate Tampa area. The citizens of Tampa had applied through the State Legislature to the federal government for 160 acres of land to be used as a county site for Hillsborough County. In this effort they were successful and Jackson was awarded the contract for the survey of this land. On October 30, 1848, he informed Butler that he had completed the fieldwork of this important survey and would file the returns as soon as possible.[13] Thus, the two most important public surveys of early Tampa's history were conducted by one of her own citizens, John Jackson.

There was a change of administrations in Washington in 1849 and this also meant a change in Surveyor General in Tallahassee. The new man on the job was Major Benjamin, Putnam, one of the leading citizens in East Florida and a prominent Jacksonville attorney. Putnam immediately questioned the survey Jackson made of the county site. In a detailed letter of August 11, 1849, Jackson justified his work by quoting a letter by Major L. Whiting that was, by instructions from the previous Surveyor General, to guide him in the survey.[14] His explanation appears to have persuaded Putnam and he was soon in line for more survey work.

The year 1849 stands out in Jackson's life for two other reasons. The first involved a serious outbreak of Indian trouble (or apparent trouble). This trouble began with the attack on two men in the Indian River settlement and the murders of Captain Payne and Mr. Whidden at the Kennedy and Darling store on present day Payne's Creek, near Wauchula, Florida. As Jackson informed Putnam, "The Indians have set the whole country in an uproar—people are gathering together in every neighborhood building Forts & Blockhouses in order to protect themselves. [T]his country never will be settled whilst the Indians are allowed to remain."[15] The second reason was the opening of his general store on the corner of Tampa and Washington Streets. Jackson, if writer Karl Grismer is to be believed, was fortunate to be able to accomplish this goal. In the great hurricane of 1848, Jackson's home, like most others near the water in Tampa, was destroyed and the contents of two strong boxes, which he kept at the store of Josiah Ferris, were carried away by the action of the water. Luckily, "Jackson employed two trustworthy Negroes to search for the strong boxes in the debris along the riverbanks. Both

---

[12]For Jackson's problems, see letters of July 17 and 29, 1848. *Letters and Reports*, 2, 107, 111-12.
[13]Letter of October 30, 1848. Jackson to Butler. *Letters and Reports*, 2, 115.
[14]Letter of August 11, 1849. Jackson to Putnam. *Letters and Reports*, 2, 127-29.
[15]Letter of July 30, 1849. Jackson to Putnam. *Letters and Reports*, 2, 119-20.

boxes were found, near the foot of Washington Street, with the cash still in them."[16] Jackson's store proved to be a life-long business but the Indians were an important source of anxiety until the end of the Third Seminole War.

The Indian Scare of 1849, which Jackson noted had caused such panic on the frontier, resulted in more troops being sent to the area. Along with the Indian problem, there was a growing fear of a slave rebellion if Indian troubles rose again in Florida. As a passing note to Surveyor General Putnam, Jackson observed in his August 11, 1849, letter, "P.S. Mr. Irwin was here a few days ago he was obliged to go to the Manatee again in order to muster a party to go with him to the Myacka where he left his Waggon & cos. &c when he was here he gave a pass to 2 Boys belonging to a Mr. Sanchez from St. Augustine to go home Major Morris the Commander at this place followed them and brought them back—on suspission of having inviagled a negroe of here away with them. He has them in the guard house instead of delivering them over to the civil authorities Mr. Irwin was gone before they were brought back he has not heard of it yet."[17] Major William W. Morris, then commander at Fort Brooke, with two companies of the Fourth Artillery as his sole support in the face of an Indian outbreak, was not taking any chances with Surveyor John Irwin's hirelings causing any disturbances or defecting to Billy Bowleg's camps. Having served in Florida during the Second Seminole War, he was not about to allow the large scale defections witnessed in the first year of that conflict. Major Morris would not be responsible for allowing a slave or an Indian rebellion while he was on watch duty.[18]

Indian tensions continued to build along the frontier. With the murder of a one Daniel Hubbard, the situation became so tense that rather than risk another war, Billy Bowlegs, who had been living near Lake Thonotosassa, decided to leave the vicinity of white society and retire to the relative safety of the Everglades. The departure of the Bowlegs encampment, the last was the last of the Indians in Hillsborough County until the outbreak of war in December 1855.[19]

The strong suspicion of a possible renewal of war against the remaining Seminoles was expressed frequently in John Jackson's letters of 1854-55. Writing the new Surveyor General, John Westcott, on July 1, 1854, Jackson quipped, "I can-

---

[16]Grismer, *Tampa*, 113-14.

[17]Letter of August 11, 1849. Jackson to Putnam. *Letters and Reports*, 2, 127-29.

[18]Donald L. Chamberlain, "Fort Brooke; A History," (Unpublished Masters Thesis. Florida State University, June, 1968), 125.

[19]Grismer, *Tampa*, 123.

not go to the field for some time yet until I get supply's which I sent for to N. Orleans. Unless the Indians get my scalp (which is the opinion of many in this part) you shall hear from me occasionally."[20] For the entire last quarter of 1854, while Jackson was in the field, Westcott did not hear from his old friend and fellow surveyor. Jackson opened his January 12, 1855, letter as follows, "I presume on account of my long silence that you begin to think by this time (with others of our neighbours) that King Billy has got hold of us but such is not the case as you will presently see on my reporting progress."[21] This somewhat playful attitude seems to have been necessary to survival facing the conditions of Florida's difficult frontier, Indians and all.

The 1854-55 surveying season was a very harsh one for John Jackson and his crew. It begun with the admonition of Surveyor General Westcott, "It is my wish and intention, so far as I can control the matter, to have all surveys made under my supervision to be more perfect than they have been heretofore, and executed strictly according to Law." The surveys of Florida, up to this time, had not been noted for their accuracy, but for the "careless manner" in which many had been conducted.[22] By starting in August, Jackson ran into Florida's rainy season, which complicated a survey that was already in trouble by having incorrect measurements for the township corners. On August 14, he wrote Westcott, "I have ran west on the standard line as far as Peas Creek we had some swimming through the swamp before we got as far as the bank of the creek every Pond and Prairie swamp &c are flowing over—I have taken the Chills on saturday last I presume its owing to my not being accustomed to wading waste deep in water for some years past."[23] By September 10, the rains had flooded the entire area and forced Jackson to suspend operations. To complicate matters, one of the chainmen caused a number of errors that forced the surveyor to resurvey portions of the area again under the same adverse conditions. He notified Westcott, "when I commence again (which I will as soon as the water falls) I hope to have better Chainmen."[24] To make sure the Surveyor General had a clear idea of the cause of this suspended operation, Jackson wrote, "I am very sory that I can not proceed with the work, as

---

[20]Letter of July 1, 1854. Jackson to Westcott. *Letters and Reports*, 2, 139.

[21]Letter of January 12, 1855. Jackson to Westcott. *Letters and Reports*, 2, n.p.

[22]Letter of June 1854. Westcott to Jackson. Drawer: U. S. Surveyors H-N. File: U. S. Surveyor John Jackson. Land Records and Title Section, Florida Department of Natural Resources, Tallahassee, Florida. (Hereafter: "Jackson File.")

[23]Letter of August 14, 1854. Jackson to Westcott. *Letters and Reports*, 2, 143.

[24]Letter of September 10, 1854. Jackson to Westcott. *Letters and Reports*, 2, 147-48.

you seem to be in a hurry with _____. I will loose but as little time as possible untill I try it again my men were very ill with the diarea &c & could not get them to continue."²⁵ By October 1, 1854, Jackson was writing that provisions were a problem, "I scarcely know what to do for provisions as there has not been a vessel here from New Orleans in 2 or 3 months and there is not one Barrel of Flower or Bread in the place; however I will be able to get some provisions in the country untill the steamboat arrives. She is expected about the 8th Inst."²⁶ Jackson and his crew(s) did not finish the work of his contract until February 1855.

This same survey also brought another problem to the fore. The Seminoles were watching the progress of his survey party and made some highly visible gestures to warn them not to enter the area. In one of the more telling letters written by the surveyor, he stated:

> I had a great deal to contend with in having a rough country, bad weather, and what was worse than all in trying to dispell the fears of the men—The Indians were watching our movements, ever after our crossing Charlepopka Creek and perticulary about the Big Prairie and thence to Istockpoga Lake they set the woods on fire about us frequently; I presume they thought to frighten us from going further on their Boundariy, however I was determined to go on with the work unless they were to come up and explain themselves, they tryed to keep out of sight all they could but in the end I caught one of them reconitering our camp it happened on Sunday near the S.E. corner of T. 34. R. 28 I was out examining the country and on my return as I emerged out of a spruce pine scrub I saw an Indian travelling along our line from our Camp I called to him and motioned to him to come up to me, he signed to me and stood untill I went to him I shook hands with him and asked him to our Camp he appeared very much mortified at my seeing him he came to the camp and east and smoked the pipe with me and was to return the next day with a few dressed Buckskins, when the Foxey (Sun) would be about one hour high he did not return nor did they set fire near us after _____. They have been complaining to Capt Casey that we frequently crossed their lines.²⁷

Jackson's candor indicated that he knew he was close to the twenty mile neutral area that was guaranteed by General Worth in 1842. It was the deliberate pol-

---

²⁵*Ibid*

²⁶Letter of October 1, 1854. Jackson to Westcott. *Letters and Reports*, 2, 151.

²⁷ Letter of February 7, 1855. Jackson to Westcott. *Letters and Reports*, 2, n.p.

icy of the U. S. Government, with heavy pressure from the state's officials to violate the line with surveys, who presumed that if the lands were surveyed, and the Seminoles knowing what that implied, they would see the fruitlessness of their resistance to emigration. It was a policy of "peaceful" pressure to get the Indians to remove and the U. S. Deputy Surveyors, like John Jackson, were the instruments of this policy. In December 1855, when Lieutenant George Hartsuff and his command violated the infamous "banana patch" of Billy Bowlegs, they were not in the area as surveyors in the manner of Jackson, Irwin and others. They were on a scouting mission for the U. S. Army seeking to locate Indian settlements and fields. There was a marked difference between the activities of Jackson and those of Hartsuff, which explains why, of the three U. S. Deputy Surveyors in the field at the outbreak of hostilities, none of them were harmed in any way. The Indians knew the difference in the functions of the groups violating their boundary, even if historians have confused the issue.[28]

Jackson's experience with the Indians in the field, immediately prior to and during the Third Seminole Wars, illustrates the dangers to which the surveyors were subject to on the volatile frontier. Yet even at the end of the war, the tensions had not totally subsided. On February 20, 1858, Jackson reported to Surveyor General F. L. Dancy:

> [O]n the 9th of this month near the station of Fish eating Creek between there and Fort Denaud 2 Indians met my waggon and made signs to my camp man to leave the prairie he was a negroe man and was so much frightened that he put out for Fort Denaud the next morning he met another Indian who stoped him on the Road and inquired for his Master he told him I was coming after him he also held up two fingers and Struck the man on the breast and signed him to be off the negroe was nearly frightened out of his wits [O]n Saying his master was coming after him the Indian got excited and struck himself on the breast at the same time pointing to the Hammock saying "a heap" I presume meaning there were a heap such as he was to meet the Master [h]e told the man to stop and put out for the Hammock but as soon as the Road was clear the negroe put whip to the mules and made himself scarce as fast as he could [After going without food and fire, Jackson crew continued and searched for their campman.] [w]e pushed on after

---

[28] Joe Knetsch, "John Westcott and the Coming of the Third Seminole War: Perspectives from Within." Paper presented to the Florida Historical Society's Annual Meeting. Tampa, Florida, May 12, 1990.

the waggon untill after about 1 Ocl when we saw some Indians ahead of us on the Road going the same way that we were going they stoped on the Road where there were some Cabbage trees extending to a hammock on each side of the Road as we approached to about 3 or 4 hundred yards of them they squatted and we could see them extending toward the Hammock on each side and every one taking a tree. I did not like the movements of the Indians and did not deep it prudent to aproch them in that position upon which we made a circuit round and came in on the Road out the other side of the Hammock my idea for so doing was that they thought we were armed with revolvers and altho' they must have been 3 or 4 times our number they would not wish to attack us openly

Jackson's crew was, in fact, unarmed and could not have offered resistance to the Indians. The campman was found the next day "crying like a child" because of the fright. Jackson sent him to Fort Meade and continued his survey, but was continually watched and having the woods set on fire around him.[29]

During the 1855 surveying season, Jackson found himself involved in the surveying of lands around Tobacco Bluff and Terra Ceia, in particular, the permits of some of the Armed Occupation Act settlers. He had been contacted early in 1855 to prepare to survey Tobacco Bluff and had even discussed this with Westcott. However, it does not appear that Westcott was in a hurry to have the area surveyed.[30] On June 11, 1855, Jackson disclosed to Westcott that he needed to know about the survey because "I have been indiscreet in mentioning the conversation that you and I had on the subject."[31] Jackson was probably correct in assuming that he had been indiscreet, he did not get the contract to survey this part of the area until 1858.[32]

Directly related to the surveys of this island area was the survey of the Joseph Atzeroth permit. This survey was very unique in that, though technically not difficult, it had a bureaucratic history that caused serious delays in Atzeroth finally obtaining his patent. In her article, "The Joseph Atzeroth Family: Manatee County Pioneers," Cathy Bayless Slusser made a special point of showing some of the difficulties faced by this early and important settler. She correctly notes that Atzeroth received Permit No. 949, dated October 29, 1844, for the land in U. S.

---

[29]Letter of February 20, 1858. Jackson to F. L. Dancy. *Letters and Reports*, 3, 3.

[30]Letters of March 30 and April 25, 1855. Jackson to Westcott. *Letters and Reports*, 2, 159-160, and *Miscellaneous Letters to Surveyor General*, Volume 2, 1849-56, 537.

[31]Letter of June 11, 1855. Jackson to Westcott. *Letters and Reports*, 2, 163.

[32]Letter of June 29, 1867. Jackson to Hugh Corley. Jackson File.

Government Lot No. 1, in Section 34, Township 33 South, Range 17 East. He, indeed, did travel to Newnansville to finish the proceedings and file additional documents, a cumbersome requirement of the law, until changed at the request of David Levy Yulee, and probably assumed things were fine. According to Slusser's research, in January 1849, the problem of mixed Permit numbers was allegedly solved by the testimony of Judge Simon Turman and Samuel Bishop. Why then, didn't Joseph Atzeroth get his patent to the land he had obviously settled until April 14, 1870? Slusser assumed that the mix-up over the numbers and the intervening war years were the causes of delay.[33] This is true as far as it goes, but it goes much further.

What Jackson's letters to the Surveyor General and others show is that there were technical problems with the survey of the grant. As noted above, Jackson was hired to survey the Terra Ceia site in 1858. The survey was not accepted by the Commissioner of the General Land Office because the starting point of the survey was not sufficiently clear. In a letter to the Register and Receiver of the Tampa Land Office, dated October 24, 1859, Commissioner S. A. Smith wrote, "The testimony in question is not sufficiently clear upon the point at issue to justify this office in concurring in your joint opinion in the case." The letter also noted that "a Stake or Blazed Tree bearing N.W." was not clear enough to establish a proper corner. Smith further questioned as to where this alleged point fell in relation to the official public surveys. All in all Smith did not feel justified in approving the patent until "competent testimony" was offered to properly establish the corner.[34] What happened next is of note in the history of Tampa Bay. The case was turned over to the Tampa Land Office for further work just as the War Between the States commenced. In a hand written note at the end of a letter from Acting Commissioner of the General Land Office, Joseph Wilson, dated May 24, 1859, is the wording "filed by John Darling in the Tampa Land Office on October 11, 1861," well after the start of the conflict.[35] As many of Darling's personal papers were burned during the war, it may be that the Atzeroth claim went up in these same flames.

However John Jackson was not a man to let a neighbor down. On June 29th,

---

[33]Cathy B. Slusser, "The Joseph Atzeroth Family: Manatee County Pioneers," *Tampa Bay History*, 4 (Fall/Winter 1982), 20-34.

[34]Letter of October 24, 1859. Smith to the Register and Receiver, *Tampa Comr's Letters*, Volume 16. Unnumbered, P. K. Yonge Library of Florida History, University of Florida, Gainesville, Florida. Thanks to Elizabeth Alexander, the individual land office letters are preserved in good condition and are an invaluable source for anyone interested in local land conditions, sales, patents, etc. The individual land offices were: Tallahassee, Tampa, Newnansville, St. Augustine and Gainesville.

[35]Copy of a letter dated May 24, 1859. Wilson to Atzeroth. Jackson File.

1867, Jackson wrote to Hugh A. Corley, Register of State Lands, asking him to look into the granting of the patent to Atzeroth "which he should have had years ago." He requested that Corley look into the documents at hand in Tallahassee to see the justice of Atzeroth's claim. As Jackson pleaded, "He is one of our best Citizens and is very much injured by not having his Patent like other settlers under the Armed occupation." He also informed Corley that Atzeroth had written him to intervene as the surveyor of the land and one most knowledgeable about the boundary. Jackson followed up with another letter dated October 1, 1867, to Corley, asking that he intervene on behalf of Atzeroth with Dr. Stonelake, Register of Public Lands for the Reconstruction Government, pointing out to him the justice of Atzeroth's claim. Jackson asked him to argue most strongly that the Atzeroth's had totally complied with the provisions of the Armed Occupation Act and were living on the land.[36] Whether these entreaties on behalf of Atzeroth had the desired impact is difficult to judge, but it should be noted that Jackson stressed their compliance with the law, their citizenship and the implication that unnamed "interested parties" were trying to oust them as arguments for the patent.[37] This last allegation was sure to set well with Stonelake and other Reconstruction bureaucrats in that it was almost universally assumed that these "interested parties" were probably unrecontructed rebels and obstructionists.

Jackson continued to have an interest in his surveying career long after he had quit the fields and settled in as a full-time businessman and service in the Confederate army.[38] He often wrote letters to the Surveyor General suggesting corrections to surveying problems that arose in his area and referred to his days as a United States Deputy Surveyor. His case was strong in asserting his position, as he was often employed by Surveyor General F. L. Dancy as an examiner of other surveyor's work. His widely recognized abilities as a surveyor, community leader and businessman assured him the continuing respect of his peers and the community as a whole. Jackson's life was full of adventure, daring, hard work and the true pioneering spirit that helped to settle the wilderness of Florida.

---

[36]Letter of October 1, 1867. Jackson to Corley. Jackson File.

[37]Letter of June 29, 1867. Jackson to Corley. Jackson File.

[38]Janet Snyder Matthews, *Edge of Wilderness: A Settlement History of Manatee River and Sarasota Bay, 1528-1885* (Tulsa, Oklahoma: Caprine Press, 1983), 254. Ms. Matthews' excellent research turned up the fact that he was listed as a soldier in the Confederate Army, but I have not found the exact unit. It is very possible he served in the Tampa area throughout the war.

To the Register of the Land Office at Newnansville E. F.

Under the provisions of the Act of Congress approved on the 4th day of August, A. D. 1842, entitled "An act to provide for the armed occupation and settlement of the unsettled part of the Peninsula of East Florida."

To all whom it may concern:

NOTICE is hereby given that under the provisions of the act of Congress above cited, I, John Jackson do hereby apply to the Register of the proper Land Office for a PERMIT to settle upon *One hundred and sixty acres* of unappropriated public land, lying south of the line dividing townships numbered *nine* and *ten*, south of the base line, and situated as herein described.

I aver that I am a single man over Eighteen years of age and able to bear arms and that I became a resident of Florida in the month of July, in the year Eighteen hundred and forty three

I aver that the settlement herein intended is not "within two miles of any permanent military post of the United States, established and garrisoned," at the time of such settlement, and that the same is not known or believed to interfere with any private claim that has been duly filed with any of the Boards of Commissioners, surveyed or unsurveyed, confirmed or unconfirmed.

DESCRIPTION OF THE INTENDED SETTLEMENT.

Range 17 Township 34 South, South East Quarter of Section No 2. Number Two

I swear that the above statement is true to the best of my knowledge and belief

John Jackson

We Certify that the above is a true copy of the original now on file in this office

Sam Russell
Register

Jno Parsons
Receiver

CHAPTER 7

D. A. SPAULDING IN FLORIDA: A SPECIAL MAN WITH EXCEPTIONAL TALENT

with Don Sonneson

The name, Don Alonzo Spaulding, must have had a magical ring to the inhabitants of St. Augustine when he arrived to work for the Surveyor General's Office in late 1854. A man of meticulous habits and strong principles, Spaulding had been asked by Surveyor General John Westcott to come to Florida and make some sense out of the mish-mash of Spanish Land Claims and Grants that existed in his office. At the time of his recruitment Spaulding was working in the General Land Office in Washington preparing General Instructions for Surveyors General, creating forms for the proper descriptions of corners for Registers and Receivers and other administrative details necessary to the Commissioner. In 1853 he had completed a short term as the Surveyor General for the District of Illinois and Missouri. Prior to that appointment, he had served as the Chief Clerk in the same office for four years. Thus, to Westcott, Don Alonzo Spaulding appeared to be the perfect man for the job.

D. A. Spaulding's bearing seemed to exude authority. From a very young age he had exhibited an interest in mathematics and surveying and studied both at the local academy near Castleton, Vermont, where he had been born in 1797. He obviously had some experience in surveying while in Vermont for when he migrated down the Ohio River and landed at Massac, Illinois, he had his instruments with him. He immediately went to work laying out the new county seat for Johnson County, Illinois, for which he received twenty-five dollars. With this grubstake in his hands he arrived in Kaskaskia, Illinois, in July of 1818 and soon joined a survey team headed north to work on a tract of land thirty miles north of

Alton. Like many frontier surveyors he survived by surveying, teaching school and holding other jobs when these two did not pan out. He was twice elected constable for Goshen Township in 1819 and 1821, respectively. In 1825 the people of Madison County elected him county surveyor, a post he held for a decade. By 1834 his skills were widely recognized and he was appointed U. S. Deputy Surveyor in charge of surveying the 3rd Principal Meridian and running, in all, eighteen hundred miles of lines in eighteen months. Over the course of his career he ran an estimated seven thousand miles of lines for the U. S. government.

Because Illinois was settled under both the French and English systems, Spaulding had a familiarity with land grants. In addition, he had executed a contract that segregated twenty-four Indian reservations out from the public lands on both sides of the Kankakee River. These demanding surveys and the amount of knowledge and experience he had running the lines of public surveys connecting all of these grants and reservations gave him an excellent background to handle the organizing of the grants in Westcott's office when he assumed that position. A man of meticulous habits Spaulding was perfect for the job.

Upon his arrival in St. Augustine Spaulding probably took up residence in one of the many inns. In the days before Henry Flagler's magnificent hotels this was the common practice of visitors and part time residents. Westcott had obtained permission from the Quartermaster General of the Army, Thomas Jesup, to use some of the vacant rooms in the St. Francis Barracks for the official offices of the Surveyor General. It was here that D. A. Spaulding came to work for the inventive Westcott. According to the "Register of Officers and Agents…1855," Spaulding was hired as a clerk in the office of the Surveyor General at a salary of $1,600 per year, the same amount as the Chief Clerk A. J. Miller. Also serving in the capacity of clerk were George Bunker, J. J. Daniel and J. Mickler, all of whom became Deputy Surveyors in Florida. The draughtsman in the office was the equally meticulous John Dick. Although the justification for paying Spaulding so much must have been his responsibilities concerning the organization of the grants and his other duties to Westcott, he was worth every penny.

Conditions in the old barracks were a little bit crowded at times as Westcott's staff numbered an even dozen men. This was twice the number of men employed by his predecessor Benjamin Putnam, who had rented rooms elsewhere in the city. Westcott justified his hiring so many staff members by noting that the workload was much greater relative to Swamp and Overflowed lands, private land claims, the advent of railroad grants and the growth of the demand for surveys after the

Second Seminole War and the Indian Scare of 1849-50. The influx of new settlers might have required this increase in staff size, but some of it was undoubtedly caused by Westcott's penchant for going into the field to examine the work of his deputies personally. He needed to have a competent staff ready to handle the workload when he was away on these inspections. In the persons of Spaulding, Miller and Dick, he seems to have had just what he needed.

Spaulding was soon given the task of locating one of the most controversial claims on the ground. The survey would be of the Joseph Hernandez Grant on both sides of the St. Johns River in Township 11 South, Range 26 East. This grant had been surveyed before in two separate surveys, on the East side by A. M. Randolph and on the West by Alexander McKay both during the 1849-50 surveying season. Each side of the river was to contain five thousand acres and had been originally surveyed by Andres Burgevin in 1821. General Hernandez, Florida's first delegate to Congress and a leader in Territorial Florida, was with Burgevin when he made the survey and verified the original work. Each corner was distinctly marked so that there was supposed to be no mistakes. McKay's work was thrown out by the United States District Court because it followed the true meridian "when it should have been run by the Magnetic Meridian." The same court threw out the Randolph survey because it did not conform to the acreage or location called for in the Supreme Court decision. Randolph assumed the improper location for ending the survey and there was some confusion as to which creek on the St. Johns was to be used as the ending point. Randolph noted in his reply to Westcott that he did not feel authorized to go beyond the creek at the northeastern corner and therefore gave him the best alternative. The deputy surveyor also noted that General Hernandez had been given an opportunity to review the survey by Surveyor General Putnam and could not, at that time, find reasonable fault with it. Because both of these surveys were dismissed as improper, Westcott investigated the causes and reported to the Commissioner of the General Land Office that a resurvey was called for by the courts. Reluctantly Commissioner John Wilson gave the permission to hire a new deputy surveyor to complete the task as mandated by the courts. Westcott did not have to look far to find the best man for the job, Don Alonzo Spaulding.

Because the survey would interfere with two other grants holders in the area, legally represented by George Fairbanks and David Levy Yulee, Westcott made doubly sure that these gentlemen were informed about the possible change in their grant's boundaries and sought their consent to the proceedings. He then drew up special instructions that included all of the ties that would relate to the new boundaries and the public land surveys of the area. This was done for both sides of the river and also demonstrated for Spaulding where the earlier surveys had erred according to the courts. The instructions called for Spaulding to follow as truly as possible the Burgevin lines and included a copy of that surveyor's work as his guide. If the original marks could not be found Spaulding was ordered to make a diligent search and should he not find any of the marks on the ground he must take the best evidence that could be obtained "on the spot, or elsewhere." Natural calls and "well established calls of the claim" were to be used assuming they were part of the original evidence of title. All of the nearest section corners outside of the grant were to be used in laying out the claim along with all intersections noted in the field notes so that all will know the location of these lines. Spaulding was to include detailed information in his notes not often found in other Florida instructions. In addition to the soil types, roads, trails, streams, ponds, creeks, etc. he was also to include, "all works of art, Houses Mounds, Fortifications, Embankments, Ditches, &c. So that the plat when constructed from them will present as far as possible a complete topographical description of the land embraced within the lines of the survey, as well as where the lines intersect these objects." Westcott also asked that he make special notation of which lands were subject to inundation and without artificial means of drainage were unfit for cultivation and the depth of said inundation, "as determined from indications on the trees, and its frequency of inundation from the best information to be obtained." This, the instructions noted, "is required that the swamp lands granted to the state September 28, 1850 may be accounted & more easily separated from the lands of the United States." Don Alonzo signed the contract on 4 June 1855 and took the field on 18 June. For his work he was to be paid ten dollars per mile for the private claim and five dollars per mile for every mile of connections actually resurveyed, traced and measured. The relatively high pay for the claim indicates the importance of the grant survey and the General Land Office's anxiety concerning the on-going problem of surveying correctly the Spanish claims in Florida.

Spaulding's work was very good and yet not without some minor controversy. In making out the northern line of the grant, Spaulding had crossed the traverse of Dunn's Creek made by Randolph that gave Hernandez a portion of Murphy's Island in Section 34 of Township 10 South, Range 26 East. This created a fractional section that had to be "harmonized" by an additional survey, presumably scrap work of a later date. Westcott also noted in his letter to the GLO of July 28,

1855, that Spaulding's work would have to be submitted to the U. S. District Court so that the claimant and the court would be satisfied with the final product. His work passed muster during the very next session of the court.

Don Alonzo Spaulding also did some private surveying for Westcott and his co-investors in the St. Johns Railway. This internal improvement was laid out in 1856-57 by Spaulding, not Westcott, and the route, with minor adjustments, was that used by this pioneering railroad. Prior to the Civil War, this railway was drawn not be an engine but by mules and/or horses. It was the butt of numerous jokes until purchased after the war by William Astor, who improved the line, brought in a small steam locomotive and ran the line somewhat successfully for a number of years. In a later manifestation the line became the property of Henry Flagler's system. The original route was laid out by D. A. Spaulding, a surveyor with exceptional talent.

Spaulding made many friends while serving in Florida including John Westcott and A. J. Miller. Miller informed Spaulding of the deaths of some of their mutual friends, including the wife of Colonel Rogers of the Florida Volunteers and Rafe Fontane, a member of one of St. Augustine's oldest families. His aid was solicited in attempting to get a seminary of higher learning placed in St. Augustine and located in the old government buildings, especially St. Francis Barracks. One of his closest friends and associates was Richard F. Floyd, who had served as the draughtsman under Putnam's reign in the Surveyor's General Office. Both had worked on the St. Johns Railway and Floyd had actually created the map of the route from Spaulding's notes and draft. A number of letters passed between these two gentlemen prior to the Civil War and most concerned the railway, Westcott and social happenings in St. Augustine. It appears from the little evidence remaining that these two friends remained in contact once the war began.

By mid-1857 Don Alonzo Spaulding was back in the arms of his family near Alton, Illinois. There he joined his second wife Sarah and his two sons, Henry and Don Alonzo, Jr., and one daughter Helen who had married the talented and well-to-do Andrew Hawley. The 1860 Census showed Don Alonzo Spaulding was sixty-three years of age and listed himself as a farmer with real estate worth $10,000 and personal wealth of $1,000, a comfortable level in frontier Illinois. He appears to have lived out the rest of his life surrounded by his family earning a pleasant living from his farm, real estate investments and occasional surveying. In fact, the 1880 Census showed the 83-year-old Spaulding listing himself as a surveyor. He died in mid-January 1891 at the age of ninety-four. His obituary noted the following attribute, "He was a civil engineer and his surveys are the standard at the present time."

CHAPTER 8

SURVEYS AND SURVEYORS OF SOUTHWESTERN FLORIDA

The history of surveying in Florida begins with the creation of the Territory and the establishment of the office of the Surveyor General for Florida. The first such officer was Colonel Robert Butler of Tennessee, the former ward of General Andrew Jackson and the son of Colonel Thomas Butler who died in a New Orleans' after blowing the whistle on General James Wilkinson's involvement in the Burr conspiracy. None of the surveys of Florida during the period between the Territorial period, 1821-1845, and the early Statehood era, 1845-1861, covered southwestern Florida. Two U. S. Deputy Surveyors did reach the area prior to the outbreak of the War Between the States. One, John Irwin, left no record of his work other than his field notes, but he did notice the havoc wrought on Tampa by the storm of April 1850, the same storm that damaged the Kennedy-Darling store on Charlotte Harbor.[1] The other surveyor to reach the area, in its eastern extremity, was Ramon Canova of St. Augustine. His surveys of the area show the extreme swampy nature of the land. Canova, who had a large contract, gave up his efforts when he received the news of the outbreak of war. He left the field and joined the Confederate cause.[2]

Only one other surveyor reached the shores of the Caloosahatchee River and his assignment was the area north of that landmark. John Jackson, a native of Ireland, was surveying the area at the time of the Third Seminole War, 1855-58, and

---

[1]*Letters and Reports to Surveyor General*, Volume 2, 1848-56, 97. Land Records and Title Section, Division of State Lands, Florida Department of Environmental Protection, Tallahassee, Florida. Letter of April 11, 1850. Irwin to Benjamin Putnam. Hereafter *Letters and Reports*, date and correspondents.
[2]*Letters and Reports*, Volume 3, 1857-61. Page number unreadable on copy. Canova to F. L. Dancy.

his crew was in constant danger from the Seminoles. However, with luck and the reluctance of the Indians and their allies to attack such crews, knowing them to be lightly armed, if at all, Jackson's crew escaped death. They did, of course, run into Indians and had to avoid direct contact. One of the crew, the Negro camp man, was stopped by the Indians and asked where his master was. Scared, but still with some wits about him, he pointed in the wrong direction and spurred his wagon on to the banks of Fisheating Creek where the crew found him the next day sobbing, scared and in extreme mental anguish. Jackson, who was an excellent surveyor and an honorable man, drafted a sketch of the area that remains one of the most accurate projections of the area at that time.[3]

With the close of the Civil War, the task of settling the wilderness of southwestern Florida began in earnest. The office of Surveyor General was reestablished under the direction of Marcellus Stearns, a one-armed Union Army Officer who later became a governor of Florida. Stearns, and his brother Timothy, who surveyed in southwestern Florida after his brother left the position of Surveyor General, was born in Center Lovell, Maine, the son of a distinguished family of Revolutionary War fame. His education included a stint at Waterville Academy and then at Waterville College, now Colby College. Here he was described as, "one of those frank, cordial, genial, open-hearted, whole-souled fellows whom everybody likes to meet—a man of integrity always ready for honest work." Yet in early 1861 his future, along with that of many young men of the day, lay with the fortunes of war. He joined the 12th Maine, Company E, as a private. Participation in several skirmishes resulted in his promotion to the rank of lieutenant and further action in and around Port Hudson, Department of the Gulf. He later was transferred to the Army of the Potomac, where, at the Battle of Winchester, he bravely raised the spirit of his troops when all of the superior officers had been killed or wounded. In the charge of his unit under his direct command, he was severely wounded and lost his right arm. Two things came from this injury, a decision to enter the study of law under Judge Josiah Drummund in Portland, Maine, and an appointment to the newly created Bureau of Refugees, Freedmen and Abandoned Claims. This latter appointment brought him to Quincy, Florida, in 1866. Stearns became very active in Republican politics and was rewarded

---

[3] Joe Knetsch, "A Surveyor's Life: John Jackson in South Florida," *Sunland Tribune*, Tampa Historical Society, Tampa, Florida, 1992. In this article the author has published a complete life of Jackson showing the importance of his work, including the laying out of the streets of modern Tampa, serving as its mayor and being the proprietor of a notable store, where many shopped for their dry goods in Tampa's pioneer days.

with an appointment to the office of Surveyor General, which he held until 1872. He brought with him his brother, Timothy and other members of his family, all of whom eventually served in the Surveyor General's office in one capacity or another.

The men surrounding Stearns were all staunch Republicans, especially Horatio Jenkins, Jr. and J. W. Childs, both of whom received surveying contracts under Stearns. Samuel Hamblin, a neighbor of Stearns in Quincy and a strong local supporter of the Surveyor General, also received a contract. Another officer in the Freedman's Bureau, William Lee Apthorp, received a contract and, later, so did his younger brother, John Apthorp. Although one surveyor, Josiah Stearns, received a contract and the Surveyor General was warned not to hire direct relatives, it was suspected that although Marcellus claimed no relationship, there may have been more than just a coincidence in name, particularly since both lived in Quincy. Yet, it should be noted here, not all of these political appointees were bad surveyors. Indeed, Hamblin, Childs, Apthorp, the two Stearnses and James Tannehill proved to be very competent surveyors. The one true failure as a surveyor, Horatio Jenkins, normally signed his contracts in partnership with Marcellus Williams, an experienced Deputy Surveyor with close contacts with David Levy Yulee, Samuel Swann and Hugh Corley. As Reconstruction was a time of active political campaigns, it is not surprising to see most of the Deputy Surveyors so blatantly political in their private lives.[4]

The man who laid out the exterior lines of many townships in southwestern Florida, including the area of modern Bonita Springs, was the well-known cartographer, William Lee Apthorp. William was born on December 31, 1837, in Lee County, Iowa, the son of a Congregational preacher. His early education was begun near home and he advanced rapidly to the Denmark Academy and the "preparatory department" at Iowa College, the forerunner to the University of Iowa. In 1856, he enrolled in Amherst College and graduated in 1859, taking a short-termed teaching position in Albany, New York. He was teaching music in Kingston, New York, when the War Between the States commenced. He immediately joined E Company of the 90th New York Infantry. His enlistment ran until 1863,

---

[4]Joe Knetsch, "Marcellus L. Stearns' Report on South Florida: 1872," *Florida Surveyor*, 1999, 22-26. I briefly cover most of the surveyors mentioned in this report in the article. For the career of Marcellus Williams, see, Joe Knetsch, "A Well Connected Man: The Career of Marcellus A. Williams," *Broward Legacy*, Summer/Fall 1993. For the career of John Apthorp, see Joe Knetsch, "Surveying the Southern Tip: Impracticable Swamp, Salt Marsh and Murder," *Florida Surveyor*, November 1992.

when he mustered into the newly formed B Company of the Second South Carolina Infantry, a "Colored Unit." This unit took part in the "Third Annual Invasion of Jacksonville" and was there when the town was burned, although Apthorp maintained it was not the colored troops that caused the conflagration. He spent most of the remainder of the war in recruiting activities in South Carolina and engaged in some local scouting against the enemy. His military career carried with it a promotion to lieutenant colonel in February 1865 and further service at Jacksonville, where he was mustered out of the service in 1866. He was a close personal friend of the Hawks family and sometimes accompanied Esther Hill Hawks on her travels in and around the Jacksonville area. Like the founders of Port Orange, Apthorp was a strong abolitionist, as was his wife, Charlotte Childs Apthorp. Both were fast friends with the Beechers.[5]

Apthorp's service in the Freedman's Bureau led him to Quincy where he may have met Marcellus Stearns. By 1868, he had moved south to Hillsborough County where he accepted a county Judgeship and an assistant postmaster's position. These political plums undoubtedly helped to pay the bills in the frontier environment of Tampa. Returning to northern Florida, Apthorp signed a contract for the surveying of the area south of the Caloosahatchee River on December 23, 1871. The instructions given to him in this contract indicated the unknown nature of the area to be surveyed, even at this late date:

> Survey, measure and mark a Standard Meridian Line beginning at the intersection of some Range Line with the Caloosahatchee River, and running South as far as practicable, the location of this line to be determined by the Deputy according to the nature of the Country, so as to give the longest and least obstructed line. <u>Also</u> a correction Parallel, beginning at the point on the Meridian marked for the South boundary of Township 46 South, and running East and West as far as practicable, <u>Provided</u> that the location of this line may be changed one township north or south, should the nature of the country require it. Said Merid-

---

[5]See the introduction to the Diary and other papers of William Lee Apthorp, Museum of South Florida History, Miami, Florida. For the Apthorp's relationship to the Hawks and Beechers, see, Gerald Schwartz, Editor, *A Woman Doctor's Civil War: Esther Hill Hawks' Diary* (Columbia: University of South Carolina Press, 1984), 175, 189, 238. Dr. Hawks served as a surgeon to the 2nd South Carolina while they were stationed in Beaufort, South Carolina. She was a leading advocate of education for African-Americans and founded many temporary schools in Florida during her service here. Colonel T. W. Osborne, another of the Reconstruction politicians and Freedman's Bureau personnel, was also a frequent companion of the Hawks and Apthorps.

ian and Correction Lines amounting by estimation to one hundred and twenty miles. <u>Also</u> the Exterior lines of Townships, proceeding in regular order East and West from the Meridian Line as far as practicable, and southward from the Caloosahatchee River, until they shall amount to five hundred and sixty three miles. <u>Provided</u> that if the actual number of miles of the Meridian and Correction Lines should fall short of the estimate, the deficiency may be made up by addition to the Township lines.[6]

These directions left much to the deputy to decide in the field. They also indicated that, in spite of the fact that the area was heavily mapped by the United States Army during the Third Seminole War, the land was still virtually unknown. Surveyor General Stearns must have had a great deal of faith in Apthorp's ability to place such responsibility in his hands.

In late 1872 when Myron H. Clay received a contract to survey the area, he was directed to run the section lines—mark the subdivisions—within the exterior lines run by Apthorp and to run the meanders of the fractional sections. Clay was also ordered to run, "such of the exterior lines as may be unsurveyed or obliterated, of the following townships."[7] This procedure was a bit unusual because the general policy was stated in no uncertain terms shortly thereafter, not to let a surveyor run the exterior lines of townships he was assigned to subdivide into sections. The reason for this prohibition was simple, the corners established on the exterior lines were to be connected by the interior section lines and this served as a check on the lines run by the person assigned the exterior lines. To allow the same person run the subdivision would potentially lead to false returns of surveys with no field checks. Apthorp, according to a letter dated October 1, 1873, may have made an error in the statement of his survey. This letter noted a discrepancy between Clay and Apthorp of about thirty chains or 1980 feet in setting the first mile post south of the river. According to Apthorp it was, "probable that the error arose either from a mistake in putting down the distance, or more probably from an accidental mark or defacement of the original 10 making it look like 40." This error, Sur-

---

[6]Contracts and Bonds: U. S. Deputy Surveyors. "Wm. L. Apthorp: U. S. Deputy Surveyor," Land Records and Title Section, Division of State Lands, Florida Department of Environmental Protection, Tallahassee, Florida.

[7]Contracts and Bonds: U. S. Deputy Surveyors. "M. H. Clay: Deputy Surveyor," Land Records and Title Section, Division of State Lands, Florida Department of Environmental Protection, Tallahassee, Florida.

veyor General G. W. Gilbert noted, would carry on to all surveys which joined the river or the old survey itself.[8] It is interesting to note that one of the clerks in the office of the Surveyor General was Anna W. Apthorp, the other clerk was Timothy S. Stearns. If this error had not been brought to the attention of the Surveyor General, many more mistakes and errors would have caused even more confusion than those attributed to the deputy surveyors today.

William Lee Apthorp continued to work in the office of the Surveyor General until he left for the North in 1877 for health reasons. He did have one additional surveying contract, that of the Daniel Hurlburt Grant in 1874. He also served as chief clerk in the office until his resignation in 1877. In the prior year, he had brought to the attention of the Board of Trustees of the Internal Improvement Fund the poor condition of the township plats in the office of the Commissioner of Lands and Immigration. Apthorp was given a contract to make as many new maps as possible, not to exceed one thousand for $2.50 each.[9] It was this contract that led to the creation of the famous Apthorp Map of Florida in 1877. The basis of the map was to replace the worn, torn maps in the office of the Commissioner of Lands and Immigration. Unfortunately, William Lee Apthorp did not live long after the completion of the project that has given him lasting fame. He died on January 24, 1879, in Springfield, New Jersey.[10]

Myron H. Clay was known to the members of the Stearns family and he arrived in Florida about the time they controlled the office of Surveyor General. He was hired as a draughtsman in the office and served there until his resignation on November 26, 1872.[11] He received his contract on January 6, 1873, to survey the lands in southwestern Florida, and soon left for the field. His familiarity with the system of surveying rendered no special instructions necessary for this veteran. When he returned to Tallahassee, he was not a well man. Within a few weeks of filing his reports and revising his notes, he died. The discrepancy he found in the Apthorp lines was duly reported, but there was no way to discuss the matter fur-

---

[8]*Letters of Surveyor General*, Volume 11, 1869-81. 335-36. Land Records and Title Section, Division of State Lands, Florida Department of Environmental Protection, Tallahassee, Florida.

[9]*Minutes of the Board of Trustees of the Internal Improvement Fund of the State of Florida*, Volume II (Tallahassee: J. H. Hilson, 1904), 141.

[10]"Introduction to the Diary of William Lee Apthorp."

[11]Miscellaneous Letters to Surveyor General, Volume 4, 1869-74. 177. Land Records and Title Section, Division of State Lands, Florida Department of Environmental Protection, Tallahassee, Florida.

ther.¹² What he put on the ground had to be accepted as correct and unchangeable. As Surveyor General Gilbert noted, "Mr. Clay's health, never very robust, gave way to the exposure and hardship of his tour." Like Sam Reid and others before and after him, Myron Clay gave his life for the profession he loved so much.

Probably no surveyor has caused as much trouble with flawed surveying than Horatio Jenkins Jr. Jenkins was neither a fool nor an uneducated bumpkin. Born on March 23, 1837, in Boston, Massachusetts, he was educated in the better schools of the area and later attended Yale University. He also attended Harvard Law School in nearby Cambridge. At that time it was not necessary to graduate from the school to enter the practice of law and prior to the War Between the States, Jenkins practiced law. A striking man with neat mustache and goatee and blondish hair, he entered the service shortly after the war began, as a private in the 5th Massachusetts Militia. He was soon elected lieutenant colonel of the 40th Massachusetts Infantry, transferring to the 4th Massachusetts Cavalry as a full colonel. He saw a great deal of action with these units during the Virginia campaigns. For his gallant and efficient service, he was given the rank of brevet brigadier general on March 13, 1865.¹³ In 1866, he, along with many others, migrated to Florida to improve their fortunes. Jenkins bought land in Alachua County and settled down as a gentleman cotton grower. Settling down to a "normal" life was not in the cards for this ambitious man. The call of politics was too strong and he entered the Reconstruction milieu with gusto.

Jenkins became involved in the moderate version of Reconstruction in Florida. In this he opposed the policies of the radical faction led by Jonathan C. Gibbs, Charles H. Pearce and Liberty Billings. Jenkins was a leader in the fight for a new constitution and was elected president of the second convention. After proposing a compromise that General George G. Meade was willing to accept on behalf of the government, Jenkins was elected president of the convention that drafted the Constitution of 1868. During this time, he was also elected state senator from Alachua County but had to fight for his seat when his occupancy was challenged during the first seating of the new government under this constitution. Once seated, he led the fight to impeach Governor Harrison Reed, whom Jenkins

---

¹²*Letters of Surveyor General*, Volume 11, 1869-81. 335-36.

¹³Roger D. Hunt and Jack R. Brown, *Brevet Brigadier Generals in Blue* (Gaithersburg: Olde Soldier Books, Inc., 1990), 314. The author would like to thank Dr. David J. Coles for leading him to this source.

believed was too radical and corrupt. Twice during this turbulent period Jenkins introduced bills to impeach the governor and actually during the second attempt admitted when pressed that he had no serious charges to make. Both bills failed. In these attempts to unseat Reed, Marcellus Stearns supported him. Jenkins also had an ally in Congress, the powerful Radical General Benjamin Butler, who had probably known Jenkins in Massachusetts and during the war. Jenkins used a circular from this influential leader throughout Florida in getting the constitution of 1868 adopted. Despite his leadership of the moderate faction, Senator Simon Conover and Congressman Josiah Walls, who opposed him and his Jacksonville ally attorney, Horatio Bisbee, removed Jenkins from his federal patronage job as collector of customs. At this time, 1873, Jenkins became interested in surveying as a politically appointed job. He was not to be disappointed.[14]

Jenkins began his surveying career aligned, as noted above, with the experienced Marcellus Williams of Fernandina. Williams, whose political ties were as equally impressive as Jenkins', had begun his own career as a surveyor in the late 1840s under the guidance of Arthur M. Randolph and other experienced surveyors. Williams, along with Charles Hopkins of Jacksonville, was one of the few men to return to the field as a surveyor following the war. The first contract called for the Williams-Jenkins team to survey along the eastern coast of Florida and across the Everglades to the west coast. This task, difficult in favorable times, proved to be impossible because of the low water in the Everglades during the dry surveying season. The pair swung around the peninsula of Florida and attempted to join their lines from the east, but in this they failed, because the weather had not cooperated. In the early part of 1875, Jenkins, on behalf of the team, applied for an extension of time to complete the surveying in southwestern Florida. With the approval of Surveyor General Leroy Ball, the surveys of the islands of Charlotte Harbor were completed, along with the extension of the Meridian line between Ranges 25 and 26. The base parallel connecting Townships 50 and 51 was not run because of the lateness of the season. However they did find sufficient water to float their canoes and finish the exterior lines and subdivision of Townships 48 and 49, Ranges 32, 33, and 34 South and East. They also surveyed Township 49 South, Range 25 East. This, again, was against the normal policy of not allowing

---

[14]For information on his activities during Reconstruction, the best source remains Jerrell H. Shofner, *Nor Is It Over Yet: Florida in the Era of Reconstruction, 1863-1877* (Gainesville: University Presses of Florida, 1974). I have not gone into great detail about Jenkins' rise and fall or his various alliances during this turbulent era. Jenkins is probably worthy of a biography for his role in Florida during this troubled time.

the same surveyors to subdivide sections of Townships where they had surveyed the exterior lines. On July 23, 1875, Jenkins informed Ball that the team had surveyed one other township, Township 50 South, Range 25 East. All of the other townships south of this were too wet, part of the mangroves surrounding the Thousand Islands and very difficult to survey.[15]

Jenkins' survey of the islands of Charlotte Harbor under this contract left much to be desired. Contrary to the specific instructions in the Manual of Surveying and common practice, Jenkins adopted a different method in running the meanders around the mangroves found on Sanibel, Captiva and Gasparilla Islands. According to a letter written by Captain Sam Ellis, John Jenkins Jr. of Tallahassee, a member of his uncle's crew on this occasion, told him that Horatio Jenkins used a new measuring device, three feet to the oar stroke, to calculate the distance around the islands. The younger Jenkins even knew that one island was more than a mile off on the recorded maps of the area.[16] Additionally, it is unlikely that Jenkins ever set any monuments on the ground on these islands since few have been found and confirmed as his and simply protracted his lines across these land forms. The corners were set theoretically, but not actually. This failure later caused a number of problems for surveyors that followed him in the area, particularly Albert W. Gilchrist, later governor of Florida.[17]

Jenkins was paid for the erroneous surveys, but still complained to Surveyor General Ball when he was shorted some $800. The cause of the shortage was innocent enough, a shortfall in the funding of the surveys of that year, however, because another surveyor, C. F. Smith, had been paid the full amount due him, Jenkins felt abused by the system. He blamed the chief clerk in the office, William Lee Apthorp, for the mix up, even though Apthorp had explained the situation to him. Jenkins felt he should have been paid first and Smith made to wait, especially since he had finished his work before Smith and the latter had overrun the costs of his contract lines by about $1,000. Jenkins immediately lined up political support to receive his funds from Congress, enlisting the aid of a General Farley in the cause. In his fight, he received the backing of the Commissioner of the

---

[15]Miscellaneous Letters to Surveyor General, Volume 5, 1875-77, 18, 20, 30, and 34. Land Records and Title Section, Division of State Lands, Florida Department of Environmental Protection, Tallahassee, Florida.

[16]Miscellaneous Letters to Surveyor General, Volume 22, 25. Letter of August 2, 1898. Sam Ellis to Richard Scarlett, Surveyor General of Florida. Land Records and Title Section.

[17]For the career of Albert W. Gilchrist, see Joe Knetsch, "Impossibilities Not Required: The Surveying Career of Albert W. Gilchrist," *Florida Surveyor*, December 1995.

General Land Office, whose hands were tied politically because of the controversial election of 1876.

Jenkins returned to Jacksonville in the 1870s and set up a law practice. There he represented a large number of individuals who had problems with the land offices and other governmental agencies and officials. He left Florida after the turn of the century and moved to Minnesota. He died on January 13, 1908, in Alexandria, Minnesota. His body was transferred to the Universalist Church Cemetery in Nashua, New Hampshire, where he had expressed a desire to be buried.[18]

---

[18] *Brevet-Brigadier Generals in Blue*, 314.

CHAPTER 9

MARCELLUS L. STEARNS' REPORT ON SOUTH FLORIDA: 1872

Of the many offices held by Marcellus L. Stearns in Reconstruction Florida, none has received so little attention as his tenure as Surveyor General. From 1869, when the office was recreated, until 1873 when he was sworn in as the State's Lieutenant Governor, Stearns served Florida as its Surveyor General. This was a political plum with tremendous responsibilities and the potential for making costly errors. He had to chose a whole new group of Deputy Surveyors whose work he could trust to be accurate and honest. Stearns had to recreate the land history of the State, clear titles, administer a complicated office and attempt to oversee the surveys of southern Florida. The job may have been a political appointment, but it was not a simple sinecure.

Stearns' origins were in the state of Maine, where he was born on April 29, 1839, in the hamlet of Center Lovell. His family had very distinguished Revolutionary War patriots in its line and some of his relatives stated that Marcellus closely resembled one, Major Benjamin Russell. Stearns received his higher education at Waterville College (now Colby) after attending its preparatory school, Waterville Academy. There he enjoyed the social life of the campus joining Delta Upsilon fraternity and being generally popular with his classmates, one who described the future Florida Governor as, "one of those frank, cordial, genial, open-hearted, whole-souled fellows whom everybody likes to meet—a man of integrity always ready for honest work." However, because of the times, the beginning of 1861, he did not complete his course of studies and entered Company E, Twelfth Maine Volunteer Infantry. He enlisted as a simple private.

The regiment was soon organized and Stearns was promoted to the rank of sergeant. This was quickly followed by an advance to the rank of lieutenant in June of 1862. After some preliminary training in the rudiments of military life, his unit

Marcellus Stearns
(Florida State Photographic Archives)

was sent southward to the Department of the Gulf. Ironically, he was given, for a very short time, the command of the Federal schooner *Hortense* which patrolled the waters of Louisiana's Lake Ponchartrain. After some heated military service in the campaign against Port Hudson on the Mississippi River, he was transferred to the Army of Northern Virginia. At the Battle of Winchester, now serving as a First Lieutenant and in the forefront of the fighting, he saw many of his superior officers wounded. Taking the initiative, he led the troops in a series of charges, in which he was severely wounded. For his valiant efforts he lost his right arm and was placed on reserve status.

Two activities dominated the life of Marcellus Stearns during this period. First, he began to study the law with Judge Josiah H. Drummund in Portland, Maine. The judge was an excellent teacher and encouraged Stearns in his studies, which he continued after he had begun his second activity, namely taking a position in the newly formed Bureau of Refugees, Freedmen, and Abandoned Claims. This new activity took him to Wheeling, West Virginia and then to Quincy, Florida, where he headed the local office. After six months in Quincy, he was admitted to the Florida Bar and becoming active in local Republican politics.

The logical step from the Freedmen's Bureau into the maelstrom of Republican politics was quick and important for Marcellus Stearns. He quickly rose into prominence and became a delegate to the first state-wide Republican convention in July of 1867. Early the next year, he served as a delegate to the Constitutional Convention. His political fortunes continued to rise until, in 1869 he won election to the newly constituted Florida Legislature as the representative of Gadsden County. In the following year, he was elected to speaker and became the only speaker in Florida history to preside over seven sessions of the Legislature, including extra sessions. He held this post until 1872. It is interesting to point out that he received his appointment as Surveyor General of Florida in 1869. It was not uncommon for politicians of that day to hold more than one appointment.

Stearns' life after his tour as Surveyor General is probably more important than his life prior to it. After a run at the governorship at the Republican convention in 1872, where he actually received the majority of votes on the first day, he settled for the post of lieutenant governor under the leadership of Ossian B. Hart of Jacksonville. Both men were considered political moderates and opposed to the Radical faction. Indeed, Stearns, as Speaker of the House, led the move to impeach Governor Harrison Reed and opposed the machinations of Dade County resident William Gleason. Unfortunately, Governor Hart died shortly after the legislative session of 1874 and Stearns became acting-governor. As acting-governor, Stearns was involved with the unusually delicate and complex politics of Reconstruction Florida. He was opposed by many of the former Radical faction and many of the leading Afro-American politicians, especially John Wallace, the author of the book, *Carpetbag Rule in Florida* in which Stearns is the leading villain. He also had to conduct the negotiations with the representatives of the Francis Vose estate, whose suit against the Board of Trustees of the Internal Improvement Fund, of which the governor sits has chairman, had tied up state-owned lands from being used for the benefit of promoting canals and railroads since 1872. The short-lived acting-governorship of Marcellus Stearns was anything but smooth and successful.

In 1876, Stearns ran for the governorship, but failed against the Democratic candidate George F. Drew. As happened throughout the nation in this election, the results were questioned everywhere and the campaign itself was very "tumultuous." Having been bitterly disappointed in his race for the governorship, President Hayes appointed him as United States Commissioner at Hot Springs, Arkansas. The next year, in a more pleasant set of circumstances, he married Ellen Austin Walker of Bridgewater, Massachusetts. Although he held different posts in different areas for many years, he did not relinquish his home in Quincy, Florida. He was, for example, president of the Atlantic National Bank in Atlantic, Iowa, from 1887 until ill-health forced his retirement in 1890. He also remained active in his law practice until the very end of his days. His final move was to Palatine Bridge, New York, where his wife's father was the minister of a local church. While preparing to head south for the winter of 1891, Marcellus L. Stearns expired on December 8. On his tombstone is listed what he considered his greatest accomplishments—the next to the last line begins, "US Surv-Gen."

With the above introduction to Marcellus Stearns completed and for a better understanding of the report that follows, it would be worth a short look at the Deputy Surveyors appointed by the Surveyor General to carry out the difficult task of measuring, platting and describing southern Florida. J. Angus MacDonald was a New York born surveyor, engineer, real estate promoter and part-time farmer in many areas of Florida, including Dade County in his later life. The special contract noted for MacDonald was for a survey of the Wekiva River in today's Seminole County. Samuel Hamblin was a neighbor of Stearns' from Quincy who had some previous experience in surveying in the north. William Lee Apthorp, a graduate of Amherst, was a surveyor, a clerk in the office of the Surveyor General of

Florida and a mapper. Apthorp also had some experiences in the Civil war, rising from Corporal to Lieutenant Colonel just prior to his discharge. He is best known for his 1877 map of Florida. His descendants have been active in the Historical Association of Southern Florida for many years. Charles F. Smith had some years as an engineer and practical surveyor in the north before heading to Florida. His surveys were some of the toughest to perform, especially his attempt to survey the area of today's Citrus County, where he quite frankly got lost and turned in what proved to be a contorted survey that is still in effect as the last official work in that difficult vicinity. The final surveyor noted by the following report is Marcellus Williams. Williams was born in North Carolina and migrated to Florida sometime in the 1840s. He began his career in surveying as a member of the crew of Arthur M. Randolph in 1847 and received his first contract for surveying in 1851. He is best known in Florida history as the leading name in the famous real estate firm of Williams, Swann and Corley, which handled much of the State's land business in the confused and frustrating 1870s. Of the group appointed by Stearns, Williams had the most experience surveying in Florida.

As the Surveyor General, Marcellus Stearns was required to report on the progress of the surveys carried out under his yearly appropriation. This meant a synopsis of the work done in the field by the Deputy Surveyors and the expenditures of the official staff. Each surveyor working under Stearns received his pay in dollars per mile of lines run and the contracts specified just how many miles each was estimated to be able to run in the area assigned for survey. The surveyors were required, in turn, to run the lines in the field, make out their field noted, draft a plat of the survey (although this is not the final plat filed in the land office) and submit an accounting of all expenses required to make the survey, including the wages paid to each member of the crew. Before any surveyor could be reimbursed for his expenses and paid for the work completed, he had to submit all of the above for the approval of the Surveyor General, who, if he approved these items, would send it on to the Commissioner of the General Land Office, whose staff would either approve or reject the accounts submitted. All of this work, both by the surveyor and his superiors, required time, often as much as a year to two years. Trouble in the field, e.g. snake bite, illness, etc., would require a request for an extension of time to complete the contract. Incorrect field notes or poor accounting would mandate a rejection of the work until the mistakes were corrected, again causing the surveyor much delay in getting reimbursed for his efforts and expenditures. For the surveyor to get his pay on schedule required the proper directions and support of the Surveyor General. In this aspect of the job, Marcellus Stearns appears to have been a fairly efficient administrator.

Because the majority of the remaining surveys to be run in Florida were those in the most southern region, the reports of the Surveyor Generals indicate the tasks facing the surveyors and the nature of the land which was to be measured. With the exception of MacDonald's work, all three of the deputies hired in the 1871-72 surveying season performed their duties in the southern portion of the State. The work was known beforehand to be very difficult, dangerous and, in some cases, impracticable. The land was basically that found in the Everglades, coral rock and peat bogs covered with grass and other vegetable matter. When they were not working in this type of environment, they faced true swamps and scattered cypress hammocks, called "islands" in the Everglades. The crews faced mosquitoes, snakes, flies, alligators and any other creature imaginable found in South Florida, including the now nearly extinct Florida panther. The task of surveying in this challenging environment was formidable.

What follows is the report written by Marcellus Stearns covering his offices duties and accomplishments for the year ending in mid-1872. Aside from the administrative detail found herein, the descriptions of the territory of southern Florida, the life-style of the Seminoles, from the white man's perspective, and the physical difficulties of life on the Florida frontier make for some rather interesting and informative reading. The document from which this is transcribed can be found in the Quarterly Reports of the Surveyor General's Office, Land Records and Title Section, Division of State Lands, Florida Department of Environmental Protections, located in Tallahassee, Florida.

    Surveyor General's Office
    Tallahassee Florida
    September 25th 1872
    Hon. Willis Drummond
    Commissioner General Land Office
        Washington D C
 Sir
    In compliance with your instructions of April 5th, I have the honor to submit for your consideration the following report of surveying operations in this district during the fiscal year ending June 30th 1872, together with tabular statements [not found] of office and field work.

                Surveys

All the field work undertaken during the present year has been completed and the work reported to this office, excepting the special contract of Deputy Macdonald.

Contract No. 8 made with Deputy Samuel Hamblin was the first contract for the year, the Deputy took the field about the middle of December and returned his work to this office the 16th day of May following, as his work North of the Caloosahatchee River exceeded the estimate for the whole, no work was done by him South of the River, the office work was completed and the duplicate plats and transcribed field notes, with accounts of deputy forwarded to the General Land Office on the 22nd day of July.

I contracted with Deputy Wm. Lee Apthorp on the 23rd day of December for the survey of a Standard Meridian Line, from the Caloosahatchee River South as far as practicable for the survey of a Correction Parallel between Townships 46 & 47 South running East from the Meridian Line to Lake Okechobee or to the marshes of said Lake, and that from said line to the Gulf of Mexico. Also for the survey of Township Lines embraced within the above mentioned lines until he should have run five hundred and sixty three miles of Township Lines, owing to some unavoidable delays, the deputy did not get into the field until some time in March, the season being so far advanced he failed to make his full amount of miles before the rainy season set in; which drove him from the field. His work was reported July 1st and the diagrams and transcribed field notes together with his accounts were forwarded to the General Land Office July 31st.

Contract No. 10 dated July 3d 1872 with Deputy M. A. Williams for the survey of Key Largo and adjoining Keys was reported July 6th, the Office work is now being pushed forward as rapidly as possible and will soon be completed and forwarded.

Contract No. 11 with Deputy Macdonald for a special survey has not been returned, nothing has been heard from the deputy so no reasons can be assigned for the delay.

The contract made with Deputy M. A. Williams for the survey of Township 45 South Ranges 41, 42 & 43 East and Township 46 South Range 41 East, contract dated 24th of April 1871, afterwards extended to March 1st 1872 and again to June 1st. was reported to this office June 1st. The office work was delayed for some time on account of some irregularities in the notes and as the Deputy's whereabouts at this time was uncertain no communication could be got to him. The office work is now completed and the duplicate plats and notes will be sent forward as soon as examined and approved.

Contract No. 7 with Deputy Chas. F. Smith which was to have been executed by Aug. 1st 1871 and which was extended to April 1st 1872 has not been executed under date of April 1st. I received a letter from the Deputy asking that the contract be cancelled as Mr. Westcott [a former Surveyor General of Florida], the party interested, had failed up to that time, to furnish the required data to enable him to locate the grant, though repeatedly solicited for such data. I would therefore recommend that the contract be cancelled, thereby relieving the deputy from any further responsibility in the matter.

Character of Country Surveyed

The country North of the Caloosahatchee River surveyed by Deputy Hamblin is generally pine, some small Hammocks on the River in Range 26 which he reported as very rich and susceptible of cultivation.

West of Range 25 and between the mouth of Pease Creek [River] and the Caloosahatchee River the pine lands are of good quality high and somewhat rolling, well timbered, little or no saw-palmetto, and being below the frost line are of very great value for raising tropical fruits, the soil is good and of considerable depth. Sea Island cotton was found growing wild in many places, the plant looked well and was heavily fruited.

South of the river, the coral rock comes very near the surface though there is a very large quantity of good merchantable pine growing in this region which is growing more valuable every year, the soil is suitable for pine apples and small fruits. The settlers at Fort Myers raise some Oranges, but to protect the trees from being blown over by heavy winds on account of the thinness of the soil, they place heavy timbers around them at a distance of four or five feet from the body of the tree. The Orange grown here is large and juicy and is excelled by none.

There is a considerable settlement at Fort Myers mostly interested in stock raising. The number of cattle south of the river is estimated at seventy five thousand (75,000) head. The prairie and saw grass bordering Lake Okechobee furnishes excellent pasture and is good at all seasons of the year. Cattle are shipped from here to Cuba and bring a good price. The fisheries here are entirely neglected though there is probably no point on the coast where such facilities could be obtained as here, for several months the river and bay is literally alive with schools of mullet which could be taken by the hundred barrel and when properly cured find a ready market.

On the Eastern part of his contract Deputy Apthorp found several Indian families living in small palmetto shanties, they had small patches in cultivation in corn, beans, pumpkins &c. though they seemed to subsist mostly on game and fish. Deer and other game was found in abundance and the ponds and creeks were full of fish. The Indians dress their deer skins and sell them at Forts Thompson or Myers for whiskey, tobacco and such articles of clothing as they need. There were but few of them, not more than fifteen or twenty men, women and children. They were peaceably disposed and a party of four or five of them spent several days with the deputy in his camp. They still speak their native tongue though they can understand some English. They carried an interpreter with them whom they called the Doctor and who seemed to be principal man among them. There are several small parties west of the Everglades, some near the mouth of the Kissimmee River and still others who live in and East of the Everglades. They do not seem to have any recognized chief over them though in each community one of their number is looked up to as the head of the party. There are probably not more than two or three hundred of them living in the State, and they occupy lands that would be untenable for white men, so it is fair to be supposed they will not be disturbed for years to come. Their wants are few and easily supplied. They have no difficulty in getting plenty to eat, they can erect a shelter that is all they would have it in two hours, and as for clothing in this warm climate, the less the better. They were filthy and looked healthy and must certainly be happy.

The lands surveyed by Deputy Williams at the South end of Lake Worth are of little value unless drained. On the Lake are several bodies of good hammock, not large enough however to attract settlers to that point.

The Keys surveyed by Deputy Williams have on them considerable rocky hammock very productive and seemingly very desirably located for raising fruits. The growing of pine apples on Key Largo is now an established success. Mr. Baker who cultivates them largely on this Key makes it a success financially, and the fruit is as fine as can be raised in the West Indies and more easily got to market. The timber growing on these keys is entirely different from any found in any other part of the State being principally Crab Wood, Poison Wood, Mastic, Maderia wood, Wahoo, Plum and Gumbo Limbo.

The reefs outside protect the keys from the heavy storms, and the hammock is generally high enough to be cultivated without fear of Overflow.

Under the appropriation for continuing the surveys for the present fiscal year, the remainder of the Keys from Key Largo to Key West will be surveyed and the subdivision of the country South of the Caloosahatchee River will be pushed forward to the full extent of the appropriation.

Accompanying this report are the following documents
"A" Map of the State showing the progress of surveys
"B" Report of surveying operations in this district during the fiscal year.
"C" Statement of the present condition of contracts not closed at date of last annual report.
"D" Report of plats furnished the District Land Office
"E" Report of deposits for special surveys.
"F" Estimate of Appropriations required for the Office of the Surveyor General and for continuing the Public Surveys within the district for the fiscal year ending June 30th 1874.

All of which is respectfully submitted
Very Respectfully
Your obt. Servt.
M. L. Stearns
Surveyor General

CHAPTER 10

BENJAMIN CLEMENTS

Major Benjamin Clements was one of the most important surveyors in the history of Florida. In his very first contract with Surveyor General Robert Butler, a fellow Tennessean and intimate friend of Andrew Jackson, Clements was assigned the task of laying out the Prime Meridian, which divides all surveys between east and west in Florida. He followed this contract with one for the difficult survey of the Basis Parallel, West, running from the Prime Meridian to the Perdido River, dividing the western part of the territory north and south.[1] Thus this remarkable man ran two of the most essential surveying lines utilized on all maps of Florida.

Clements came highly recommended to Robert Butler by Andrew Jackson, in whose household the young Butler was reared. Jackson's brother-in-law, General John Coffee, Surveyor General of the Territory South of Tennessee, also sent Butler his recommendation of Clements and his friend, James Exum:

> Col. James W. Exum and Major Benjamin Clements, has made known to me their intention to apply to you for appointments as deputy Surveyors in Florida, and have requested of me letters of recommendation &c., to which I most chearfully comply, and will remark to you that both of these Gentln. have surveyed for me in this department from its commencement, to its termination and I have always found them prompt, and strictly attentive to the discharge of the work assigned them, and are both of them excellent surveyors, and I have no doubt but

---

[1] Drawer, "U.S. Deputy Surveyors: A-H." File, "U.S. Deputy Surveyor, Benjamin Clements, Contracts & Bonds." Land Records and Title Section, Division of State Lands, Florida Department of Environmental Protection, Tallahassee, Florida. Hereafter, Contracts and Bonds. These valuable files contain all of the contracts and bonds of the U.S. Deputy Surveyors and frequently include important correspondence which was not filed elsewhere.

they will execute any work you may assign them, with correctness and dispatch."[2]

This letter was added to one written by John Bright, another Tennessee acquaintance of Butler's. Bright's ending sentence accurately sums up Clements' personality, "As to Major Clements abilities as a Surveyor there is no doubt, as to his correctness and integrity in other respects there is as little, he is also a man of great industry and purseverance and will if possible accomplish any thing he undertakes."[3] In the days and years ahead, this personality trait was to be greatly tested under some of the most trying circumstances found in the history of Florida surveying.

Benjamin Clements did have a great deal of experience upon which to draw prior to his arrival in Florida in 1824. He received a contract for surveying in Alabama in 1817 and continued to survey there until the public land surveys there were terminated in 1823.[4] In a letter of November 1819, Coffee informed the head of the General Land Office Josiah Meigs that Clements had been contracted to lay out the town located at Fort Jackson on the Alabama River.[5] Thus, Benjamin Clements had experience at laying out the lines for public land surveys and for platting out towns, a skill he used to advantage in 1835 when he assisted his son, Jesse, in surveying out the town of St. Augustine, Florida.

In mid-1827, Clements, in company with another Jackson relative, J. R. Donalson, left for Pensacola from Nashville, Tennessee. The crops that year were poor because of a wet and cold spring that dampened the prospects of a large yield.[6] Along the way, they visited Clements' property in Selma, Alabama, where he reported to Butler that both men were in good health and expected to reach Pensa-

---

[2]Applications for Employment, Volume 1, 1824-44. Letter of July 31, 1824. John Coffee to Robert Butler. Land Records and Title Section, Division of State Lands, Florida Department of Environmental Protection, Tallahassee, Florida. These letters are the original letters in three bound volumes. They are open to the public on microfilm only as they are very fragile and subject to damage.

[3]*Ibid*, Letter of July 25, 1824. John Bright to Robert Butler.

[4]*Territorial Papers of the United States*, Volume XVIII, *The Territory of Alabama: 1817-1819*. Clarence E. Carter, Editor (Washington: Government Printing Office, 1952), 277.

[5]*Ibid*, 732.

[6]*Letters and Reports to Surveyor General*, Volume 1, 1825-1847. Letter of June 25, 1827. Clements to Butler, 57. Land Records and Title Section, Division of State Lands, Florida Department of Environmental Protection. Tallahassee, Florida. Hereafter, Letters and Reports, Volume number, date of letter and page number. Again, these are bound copies of the originals letters and are available for inspection from the microfilm copies available at the Land Records and Title Section.

cola by early August.[7] Clements's assignment was to survey private land claims and some public lands in the Pensacola region to the amount of six hundred and fifty miles of lines. The rate of pay was to be four dollars per mile.[8] Little did he know that the costs of surveying would be much greater than the money he would realize from the completion of the contract. Nor did Clements expect to find the resistance to his attempts to survey the land claims from those most expected to benefit from such work.

The resistance to Clements and Exum in their attempt to survey the Escambia and Pensacola area land grants came from the claimants themselves. The main reason for this opposition to the surveying of the grants was the Spanish law that gave those who found errors in surveys a part of the grant or one fourth the value of said portion found in the error. Because the land commissions of East and West Florida found so many grants were either fraudulent or poorly surveyed, the recipients of the grants were leery of anyone attempting to find the exact boundaries of these plots. Should an error be found in the surveying of the grant, the surveyor could potentially become a rich man. Although this was not the case, the fear the old law engendered made many landholders reluctant to allow any survey to be made and some were hesitant in pointing out the boundaries of their grants. Butler warned Clements that some problems might exist and that he would need to exercise caution, "I congratulate you and Mr. Donalson on your arrival and hope for you every thing good; but you must put on the garb of patience _ you will have need of it."[9]

On August 9, 1827, Clements wrote to Butler stating his observation about this situation:

> I Got to this place 3d Instant in good health & have been here waiting Since have not yet received the first information from a Claimant relative to his Claim & fear I will not from what I can learn from Col. Exum as regards the dificulty he has had in procurring the necessary information & being in the neighbourhood of the Same Some of the Claimants Say, it is too far to go this hot weather & Shew there begin-

---

[7] *Letters and Reports*, Volume 1. Letter of July 29, 1827. Clements to Butler, 61.

[8] Contracts and Bonds. Benjamin Clements.

[9] "U. S. Surveyors A-H", [File] "U S. Deputy Surveyors, Clements and Exum." Letter of August 21, 1827. Butler to Clements and Donalson, 10-11. Land Records and Title Section, Division of State Lands, Florida Department of Environmental Protection, Tallahassee, Florida. Hereafter, Clements and Exum, date of letter and correspondents.

ning corners Some not got there papers at home &c &c &c I shal wait here my time out & then take the woods hot or cold if will do the best I can giving them notice that I will be in this place again in a few weeks &c."[10]

Five days later he reported that he had some of the information from nearly all of the claimants on the East side of the Escambia River, but was having some difficulty in locating the corners of the various grants. He also informed Butler that he had "taken one stand amongst them" to convince them to show him the places of beginning for their surveys and was firm in his dealings with the grant holders. If he had not done so, he believed, "I might wasted here the balance of the year."[11]

The problem of expenses constantly worried Clements on this survey because a surveyor had to perform the work at his own expense until he was paid for the completion of the survey. This process could take up to two years or more on the Florida frontier, which had no active banks that received federal funds. Almost all of the funds sent to the surveyors in the Territorial period of Florida were drawn on federal accounts in Mobile or Savannah. These funds were normally paid in drafts upon these banks, which meant that the surveyor would have to travel to Mobile or Savannah or find a local merchant who would accept these drafts as payment. Surveyors had to be men of enough means to enable themselves to attempt to fulfill their contracts or had to rely upon the ability of the Surveyor General to provide funds to them when needed, a very questionable practice in the days of unreliable mails and threats of violence to mail carriers.

This dilemma is clearly illustrated by the Clements correspondence in 1827. On the August 9, he wrote to Butler that he was short of cash because he had depended upon $500 to be sent to him from the Surveyor General. Because of this dependence he pleaded, "I have come without much cash depending on the $500 you will please Send it to my relief Soon. I have not more that $60 or $80 with me depending on that alone This is a hard Country to live in without cash &c."[12] By August 14, Clements noted, "I have been compeled [sic] to borrow $160 to enable me to go on you will please send me the necessary apparatus &c or I Shall have to send to your Man for help as came here relying on it alone & made no other prep-

---

[10]*Letters and Reports*, Letter of August 9, 1827. Clements to Butler, 65-66.

[11]*Letters and Reports*, Letter of August 14, 1827. Clements to Butler, 69-70.

[12]*Letters and Reports*, Letter of August 9, 1827. Clements to Butler, 65-66.

aration at all I cannot live here without cash."[13] By the end of the month, he had received the first letter from Butler. "I have," he confirmed, "a few minutes Since received the first line from you & find enclosed the left hand half of 5 one hundred Dollar notes the other halves I hope to get Soon &c."[14] Not until September 9, 1827, did Clements receive the right hand halves of the five one hundred dollar notes, which enabled him to carry on, but with some difficulty. As the Major informed Butler, "I also inform'd you that I have this day needed the right hand of the five bank notes of one hundred Dollars each the other halves I have receiv'd sometime before the Two hundred dollars in Georgia money I think will be of little use to me here The balance I think will go well."[15] The problem of sending money was obviously a difficult one and the fact that it was sent by halves indicates the uncertainty of the mails in the Territorial period. Also, the notation that the Georgia money would not be of use in Pensacola gives an idea of the dilemma faced by frontier settlers when they did receive bank notes. The values of these notes varied greatly and the uncertainty of a bank's solvency made handling bank issued currency a problem for both settler and sutler.

The survey of the Escambia area was more costly in another way to Benjamin Clements. On August 28, 1827, he informed the Surveyor General of some unpleasant news, "I left camp yesterday early in the day left all well when I left camp but in hour since two of my boys came in search of me & tell me that 8 of my hands was [stricken by] yellow [fever] by evening was taken Sick which is not so pleasant Since I am going to take medicine plenty with me the weather has been very hot & we have been a great deal in the Swamp &c."[16] By August 30, things had not improved:

> I am at this time in camp at Florida 20 miles above Pensacola am good health myself but dont know how long I can to remain so I have bad news to tell you on the 27th Inst. after I left home for Pensacola two of my boys were taken sick Hosea was one of the last that was taken he is very sick & three of the others one is I hope a little better two I fear dangerous without Some relief Hosea I hope not yet dangerous I have a bad chance of a Doct. I have medicines plenty with me but it does not appear to do much good yet if there is no alternative by

---

[13]*Letters and Reports*, Letter of August 14, 1827. Clements to Butler, 69-70.
[14]*Letters and Reports*, Letter of August 28, 1827. Clements to Butler, 73-75.
[15]*Letters and Reports*, Letter of September 9, 1827. Clements to Butler, 81-82.
[16]*Letters and Reports*, Letter of August 28, 1827. Clements to Butler, 73-75.

morning I Shall leave here to Pensacola for a Doct. & as soon as we can travel I shall move out from the Bay & River.[17]

As a postscript to the letter, the troubled surveyor/father noted, "Hosea has at the present a high burning fever & his nerves appear much affected & I dont know what to do for him. O, if he only was at home with his mother Such a time as this I have never seen before in all my Travels five lying here at one time bad Sick if myself & one other gets sick What will I do when no body to wait on us but I hope for the better."[18] The severe anguish of a father seeing his son, and his companions, fall sick to the yellow fever was an awful strain on the brave surveyor. On September 9, he reported to Butler that Hosea was in the hands of a Mr. Simpson, "one of the best hands in any country" and resting, but could not raise his head or turn over without aid.[19] The distraught father would not leave his son's side and had to postpone the work on the private claims. At the end of October 1827, Hosea Clements and one other member of the crew died. Butler wrote to his friend and Deputy Surveyor on November 5, "I have just received your letter from Tennessee in which you mention your return to work in last month. I do most sincerely sympathize with you for the loss you have sustained in the death of Hosea. He was a youth of much promise and one I esteemed highly."[20] The costs of the survey of the Escambia country was, indeed, very high for Benjamin Clements.

As if the death of his son, the lack of money and the resistance of the claimants were not enough, Clements had also to solve the problem of foreign measurements. What was an arpent? Did the Spanish surveyor use the Spanish or French measurement? On August 14, Clements wrote to Butler asking instructions on the Mitchel claim, since it was different from those he had previously surveyed. As he noted to the Surveyor General, "I assure you I am at a grate [sic] loss in this one case &c."[21] Butler informed his deputy, in reply to his inquiry, "for your government and that of Mr. J. R. Donelson [Donalson] to whom you will communicate, I inform you that the Arpent of France contains one hundred square poles of 18 feet each, and bears to the American acre in the proportion of 512 to 605. and the french [sic] foot as to the American as 16 to 15."[22] Clements appears to have

---

[17]*Letters and Reports*, Letter of August 30, 1827. Clements to Butler, 77-78.
[18]*Ibid*
[19]*Letters and Reports*, Letter of September 9, 1827. Clements to Butler, 81-82.
[20]Clements and Exum. Letter of November 5, 1827. Butler to Clements, 12.
[21]*Letters and Reports*, Letter of August 14, 1827. Clements to Butler, 69-70.
[22]Clements and Exum. Letter of August 21, 1827. Butler to Clements, 10-11.

been dissatisfied with his own work regarding the length of the arpent, however he did conclude:

> As respects the length of the pearch of parish Col. Exum makes 19 2/10 feet which makes to 800 Arpents 677 Acres but I am of the opinion that he is incorrect in his calculations. [A]s Such I am waiting here to see him. When I saw last he still continued of the opinion that his calculation was right & as I was of different opinion. I have been a little at a loss to know what course to pursue but on examination of my calculations at present I make 800 Arpents 677 Acres agreeable to the length you give of the French [*sic*] foot tho I may be in error if so I hope you will correct me as Soon as possible & I must go over again & make it right before I am done &c.[23]

From the available correspondence, it does not appear that Butler had too much patience with Clements' dilemma over the size of the arpent. On November 5th, in the same letter that he expressed his sympathies, he informed Clements, "I have previously given you the information sought for in relation to the calculation of Arpents and my decision on the private claims which was that you were not authorized to alter the locations of private claims when marked lines existed."[24] This quick and terse reply seems to have ended Butler's discussion of the topic, although Clements was not totally convinced and asked for further instructions on August 30. No reply to this letter has been found and it can only be assumed that the measurement, which relates 677 acres to the arpent, was correct.

Benjamin Clements also had questions regarding the location of the donations found within his surveying district. A "donation" is simply another name for a grant of land from the federal government and frequently, in the Territorial period, they approximated a section of land or 640 acres. In asking for directions about how to proceed, Clements indicated that his original instructions did not cover the relationship of donations to the section lines already run in the area. "Col," the frustrated deputy wrote, "I am at a loss to know whether or not the donation granted to a few settlers are to be run of[f] or are they confined to the Sectional lines I have 8 of them in my District. Col White tells me they are Confined to the Sectional lines & not to be run of[f] I wish your opinion on that matter as my Instructions Say nothing about it."[25] Butler responded to this question quickly

---

[23] *Letters and Reports*, Letter of August 28, 1827. Clements to Butler, 73-75.

[24] Clements and Exum. Letter of November 5, 1827. Butler to Clements, 12.

[25] *Letters and Reports*, Letter of August 14, 1827. Clements to Butler, 69-70.

and to the point, "The Donation cases you are to lay down as near to the sectional lines of the public lands as possible in one entire body so as to include all the principal improvements claimed by the proprietor, not to include more than six hundred and forty acres."[26] In other words, give the donation holder one section of land and include all of the improvements within the survey when possible.

That Clements actually went back into the field after the death of Hosea and the other young man in the face of sporadic outbreaks of the fever and extremely poor working conditions indicated the tenacity of the surveyor. Few today would begrudge him the time off from work and the request for an extension of his contract. However, because the necessary surveys were not completed and the claimants were still clamoring for satisfaction, he was required to put his loss behind him and proceed with the work. As he informed Butler, "I return'd again Early in November & took up my work I went again to Pensacola about the 22d of November for the purpose of Seeing Some person or come themselves & Show me their lands Some did So & had there land Surveyed others would not come all of which I Surveyed that I could by any means find Some with the utmost dificulty [sic] &c."[27] Yet, he still did not complete all of the private claim surveys required by the General Land Office and had to take the field again in the following year.

In 1828, Clements was again teamed with James Exum, another experienced surveyor from the Alabama Territory, to finish up the public land surveys in and around Pensacola and to complete the laying out of the private claims in West Florida. Their orders were specific:

> [Y]ou will proceed west, and take up all unfinished work above the parallel, except when it may interfere with Forbes & Innerarity's large Claims completing the same as far as practicable under your General Instructions. You will be furnished with a list of private claims which have been sanctioned and yet remain to be surveyed. Your attention is particularly called to the completion thereof so far as you can identify them. Should there not be found a sufficient quantity of Surveying to Complete your Contracts or the Season too far advanced before you Complete, you are authorized to desist from work provided you deem it hazardous to Continue the Same. It is extremely desirable that your visit to the west should supercede in future the necessity of again send-

---

[26]Clements and Exum. Letter of August 21, 1827. Butler to Clements, 10-11.
[27]*Letters and Reports*, Letter of July 10, 1828. Clements to Butler, 85-88.

ing any Surveyor in that quarter. Wishing you much health and prosperous work.[28]

Although the surveys were completed as far as they could be identified, the problem of grant locations was still a major impediment to finishing the work. Also, the concept of "sufficient quantity" needs an explanation. Simply put, the contracts for deputy surveyors usually read that the surveyor agreed to measure a certain number of miles of survey lines and not specific grants, sections or other small divisions. It was unusual for Colonel Butler to wish any surveyor "health and prosperous work" in his official correspondence; however, given the great loss suffered by Clements during the 1827 season, this inclusion indicated the concern he had for his friend and deputy.

The tone of Butler's next letter, written on May 13, 1828, however was somewhat more critical. The reason for this was the large number of criticisms he had received from the Commissioner of the General Land Office concerning the surveys of the private claims in West Florida. Although somewhat technical in nature, the criticisms were very important to the successful surveying of the land. It appears from Butler's letter that Exum and Clements did not, in every case, follow instructions strictly and did not include some of their bearings in their field notes. Sometimes they did not connect all of the surveys of private claims to the survey lines of the public lands, which created confusion for those trying to find their property. In one case cited by Butler, the township lines were not "perfectly protracted" and led to further confusion. These errors were to be corrected and the returns to the office made perfect before any payment could be authorized.[29] More importantly for the two deputies, Butler was now less confident in their work and would be more critical in the future.

By the end of June 1828, Butler was under direct fire from the General Land Office because of the errors found in the work of Clements and Exum:

> Glaring imperfections having been detected in the Surveys including private claims forwarded to the Commissioners office, and returned to this office for Correction, I must exact of you a more rigid adherence to your instructions, and a more strict and active attention to the accuracy of your work.
>
> I again reiterate in the most positive manner, that the accuracy of

---

[28]Clements and Exum. Letter of April 25, 1828. Butler to Clements and Exum, 8.
[29]Clements and Exum. Letter of May 13, 1828. Butler to Clements and Exum, 11-12.

every Survey must be tested in the field by Latitude & Departure; and in all cases, except where the error in closing be within a very limited degree, a resurvey must be made on the ground. In trying some of your Surveys by Latitude & Departure, serious errors are discovered both in regard to closing and Contents of Surveys which induce the belief that your work was not done by L & D. as you were specifically instructed to do.

The irate Surveyor General then proceeded to cite specific errors, such as the section in Township 3 South, Range 31 West, which by the Clements and Exum survey contained 1192 acres, somewhat above the expected 640. He also noted that some of the surveys of the private grants were not done according to the calls, which neither man was authorized to do.[30] The correction of these errors meant a resurvey of the area by the deputies, at their own expense. It also created an atmosphere of distrust of their work by all those required to scrutinize their work.

At the end of the year, Exum and Clements were still in the area attempting to complete their work. To the finishing of the surveys of the private claims was added the testing of the plat of Pensacola by James Exum. The plat had to be tested in order to assure the authorities that the lots granted to private individuals in the city were patented correctly and the right amount of land was granted to each individual grant holder. Any errors were to be reported to the Surveyor General's office for correction and replatting.[31]

A final problem survey was also included in this critical correspondence that demonstrates the extent of political involvement in the surveying business. This involved the island in the Escambia River claimed by Congressional Delegate Joseph White, one of the leading experts on Spanish land law in the United States. There had been some difficulty earlier with this survey in that Mr. [ ] Williams, White's agent who was instructed to point out the corners of the grant, did not show up at the appointed time. Clements wrote to Butler that, because of the large number of sloughs that ran through the grant, it would be impossible for any stranger to survey this land correctly without assistance.[32] White's insistence led to the final issuance of title to the island and the notation that it was for 800 Arpents *more or less*. In November 1828, Clements was ordered to survey this

---

[30]Clements and Exum. Letter of June 30, 1828. Butler to Clements and Exum, 14-15.

[31]Clements and Exum. Letter of November 5, 1828. Butler to Clements and Exum, 19.

[32]*Letters and Reports*, Letter of July 10, 1828. Clements to Butler, 85-88.

grant in "conformity to the Original plat."[33] White's influence in Congress appears to be the motivation for the issuance of a title to this property prior to an official survey of the land, something not in conformity to general practice.

From a technical point of view, the most difficult survey that Clements had to perform was the reconnection of the parallels. It appears from his letter of January 8, 1830, that an error was made in running the line in Range 29 West and in crossing the Escambia River. The lines coming from the west did not match with those coming from the east, run by Clements at an earlier date. As Clements pictured it:

> being instructed in the later part of the year 1825 I believe to run the line from the paralell [sic] line west of the Escambia Bay as a Meridian line to the South boundary of Genl Coffees survey and close the work East of said line to the Escambie River & Afterwards the work being continued from the Meridian of this place to the East bank of the Escambia River & not being instructed to cross the same with the survey of the lines above the parlell [sic] I am not able to say any thing the connection but Sir it is presumed it would be a mere accident if the work was to close the circumstance of the error I fear in the paralell [sic] on the 29th Range will make the Survey clash occasionally if they are now to be connected across the River &c.

Clements offered to correct these errors as soon as possible with Butler's approval, however, he noted that the work would have to be done in an area he considered impracticable to survey, since it was mostly river swamp and ponds.[34]

The survey of the Escambia country was further delayed by the high water in the river in early 1831. Clements and Exum had been sent there again to finish up all of the incomplete surveys, but, as Clements noted that even if I receded within two months, the area and time would be found to be in the midst of the sickly season and it would not be prudent to attempt the work at that time. By October, both men had their crews ready to survey in Township 1 North and Range 28 West, where local residents had several sawmills and brickyards operating in the settlement. The terrain was so rough that the survey was not finished until November 1831, when Jesse B. Clements, Benjamin's son, reported to Colonel Butler that within ten days he expected that the work on the final donations would be com-

---

[33]Clements and Exum. Letter of November 5, 1828. Butler to Clements and Exum, 19.

[34]*Letters and Reports*, Letter of January 8, 1830. Clements to Butler, 89-90.

plete.[35] One could almost hear the sighs of relief from the Clements family with the news of the completion of the surveys of the Escambia country.

Surveying the Escambia country took nearly four years. The work was difficult, dangerous and deadly. This work did not make any surveyor rich and was probably done at a cost to Clements and Exum that exceeded their pay. Above all, the losses of life that occurred during the course of the surveys could not be made up. Only the perseverance of Benjamin Clements allowed the work to be done in the most correct manner possible. The errors found did slow down the settlement of the area, especially the misconnection of the parallels, which caused the Surveyor General to suspend public sales for nearly six months in that section of the country. Yet, through it all, Clements pushed on to their final completion. The bravery, tenacity and dedication to the job shown by Benjamin Clements says volumes about the type of men who mapped the frontier in Florida and allowed the final settlement of the land to take place. Without this type of dedication, the Territory of Florida would not have evolved so rapidly toward statehood and the peopling of the land would have been delayed for an unknown period.

---

[35]*Letters and Reports*, Letter of November 24, 1831. Jesse B. Clements to Butler, 145.

CHAPTER 11

R. W. B. HODGSON AND THE ORIGINS OF THE WHITNER-ORR LINE

Ellicott's Mound has always been a source of trouble since the day of its construction. From the determination of its location three major boundary lines between Florida and Georgia have been run. But, these were not the only attempts to run the boundary line. William Cone, a member of the Georgia legislature and captain of the militia, complained in the 1817 session of the incorrectness of the line and requested his colleagues to commission a new line. Acting on this advice, the legislature of Georgia assigned Generals John Floyd, Wiley Thompson and David Blackshear to locate the headwaters of the St. Marys River, to ascertain whether or not Ellicott had placed his mound in the correct spot. They were joined by Lieutenant Daniel Burch, who later constructed the road between Pensacola and St. Augustine, and Major General E. P. Gaines. Avoiding any conflict with the local Indian inhabitants, this group searched for and found the origins of the North Fork of the St. Mary's River and the three Georgia generals concluded, "we are therefore of opinion that Mr. Ellicott and the Spanish deputation were correct in establishing on the northern branch the point of demarcation between the state of Georgia and province of East Florida."[1] This should have ended any further discussion, however, this was not to be.

Within two years of the generals' final report, Georgia Governor Rabun decided that the line was still in doubt and appointed Dr. William Greene to rerun the line. Greene's efforts were filled with frustration and accomplished only the exhaustion of his provisions and the running sixty-two and a quarter miles of line. He reported to Rabun that he was planning a continuation once his supplies and

---

[1] S. G. McLendo, "History of Georgia-Florida Boundary Line." Undated typescript. 9-12. Land Records and Title Section, Division of State Lands, Florida Department of Environmental Protection, Tallahassee, Florida.

men were replenished, but he never returned to complete the job. At this point, the new governor for Georgia, John Clark, appointed Colonel James C. Watson to complete the task. Watson ran his line from the juncture of the Flint and Chattahoochee Rivers to a point south of Ellicott's Mound. Georgia accepted this line and relinquished any claims to lands surveyed into lots south of this new boundary, but the boundary line was not finalized because of the transfer of Florida from Spain to the United States in 1821.

Soon after his office was established in 1824, the Surveyor General for Florida Robert Butler ordered another line to be run between Florida and Georgia. He chose Daniel McNeil who was already surveying in the area for the job. McNeil's line ran roughly fourteen chains north of the line set out by Watson. Butler ordered all Florida surveys to be run to the new McNeil Line, which caused some overlaps with lands already surveyed and sold in Georgia. That state held the Watson Line to be valid. Farris Cadle, the Georgia historian of surveying, noted that as a result of this difference over the state boundary, many long and acrimonious disputes took place between 1825 and 1859. Lines run by John McBride in 1827, Joel Crawford and J. Hamilton Couper in 1831 and Benjamin Whitner Sr. and Alexander Allen in 1854 were never accepted by the two states. Whitner and Allen's survey was the most accurate and had its origins in some mistakes found along the existing lines, but it was rejected. Not until Benjamin Whitner and Professor Gustavus J. Orr, of Emory College completed another survey was the line acceptable to both parties. This resolution came because both surveyors agreed to a methodology of following the arc of the great circle from the juncture at the Flint and Chattahoochee Rivers to Ellicott's Mound that allowed for only a quarter mile deviation from the Mound. If this requirement was not met, the survey would not be regarded as valid, a condition that was accepted by the legislatures of both states in 1859.[2]

But what brought about the reason for the new survey of the line in 1854, the one that ended in personal rancor and distrust? After all, it was this attempt at a line that set up the final Whitner-Orr Line. The story began in 1849 when R. W. B. Hodgson, then living in Tallahassee, received a contract to do the surveys of private claims and other scrap work. He had applied for a contract in May of that year, but had not heard from Surveyor General Benjamin Putnam. By August 27, 1849, he had heard from both Colonel Robert Hayward and A. M. Randolph, that

---

[2]*Ibid* and Farris W. Cadle, *Georgia Land Surveying: History and Law* (Athens: University of Georgia Press, 1991.) Cadle's even-handed treatment of the dispute can be found on pages 210-222.

he had been awarded a contract, but he had not received official notice of this at that time.[3] By the time the next surveying season came, Hodgson had received his notice and filed his bond with the Surveyor General's office.

Hodgson took the field filled with optimism and enthusiasm. Surveying was a profitable diversion from his normal occupation of operating a turpentine distillery and serving as a local postmaster. He soon found that there was something wrong in the layout of the lines he was attempting to run to. On August 23, 1850, he informed Richard Floyd, the draughtsman in the Surveyor General's office, of some peculiarities in the alleged lines:

> Enclosed herewith you find the Field Notes for that Portion of Frl. T. 2. R. 16. N. & E. which lies between the old Geo. line and the McNeil line. You will at once see another jog in the Geo. and Fla. Line. I have been doubly careful in determining the fact. In the first place, I ran the line west from N.E. > of this Township, and could find no line marked after I passed the field, in 2nd mile west, until near the 2nd post, _ there I found it plainly marked on Several trees, and among others the magnetic which he (McNeil) mentions - and after finding another notch or error, I ran (as far Notes) back from 1/4 Sec. corner Post in 3rd mile - and marked the line to the River which had not been done. I had with me as assistants, two men who were with McNeil, in running this particular portion of said Line _ the first Mr. Whitten _ the other Mr. Conner who put McNeil's point across the River, and was with them up the River, and saw them begin and run <u>west</u>. After I had done on the west side, I came back and began again in Mattox Field and run with greatest care to the River again, thus making myself doubly sure of being right. How such an Error could have been made I cannot imagine, it is just such a one as you know I reported on the Southern Boundary of this same Township, which rendered it still more strange: Yet so.[4]

Hodgson's letter indicated that the line near the river had never been put down or marked on the ground, as if McNeil had skipped a portion of a mile near the water's edge.

His findings were soon backed up when Floyd reported to Putnam on November 22, 1850. Upon instructions from the Surveyor General, the draughtsman reexamined the notes from McNeil's survey. Floyd discovered that McNeil "did

---

[3] Applications for Employment, Volume 2, 1845-56. 407-15. Land Records and Title Section, Division of State Lands, Florida Department of Environmental Protection, Tallahassee, Florida.

[4] *Letters and Reports to Surveyor General*, Volume 2, 1848-56, 931. Land Records and Title Section, Division of State Lands, Florida Department of Environmental Protection, Tallahassee, Florida.

not continue that line to the Suwanee River, on either side of it, the line at this point consequently remained unclosed until surveyed by Mr. Hodgson when he found that a jog existed in the line." Even more curiously, Floyd discovered that upon examination of McNeil's Field Notes, "that part of the line where the jog occurs, was not run by him, as may be seen by a copy of his field notes which relate to that line." McNeil's notes, quoted by Floyd, showed that the reason for this was the extreme high water in the Suwannee River at that time. One part of the line ended in a small lagoon, while the other side of the river had its line terminate in a swamp, adjacent to the river.[5]

The immediate response to Hodgson's discovery of the jogs in the Florida-Georgia line was a suspension of the work. This did not take the deputy surveyor by surprise and he informed Benjamin Putnam, "I regret exceedingly the suspension of the remaining portion of my work (Tp. 2 N. R 16 E.), however it is no more than what I might have expected, when I reported an Error in so important a Line as that of the Boundary between Fla. and Georgia, a line run by so old and experienced a Surveyor as D. F. McNeil, and the Error discovered by so young a man in the Public work as myself." He then went on to explain to the Surveyor General just what he had found and answered the questions being asked of him by his supervisor. Hodgson reiterated that he had two men from the McNeil survey on his crew and that Mr. Conner, noted as being 45 or 50 years of age, was with him when he crossed the Suwannee River and corrected him. "when I crossed the River he saw my cours, and remarked that 'it would miss McNeils line 1/2 a Qr', sure enough when I got out to the high pine woods no marks could be found, he (Conner) then said, 'I will shew you where McNeil started from on this side for I was wiht him, and assisted in putting him and his men across the river which was then very high,' accordingly we went N and he Shewed me within ten paces of the point where I found his (McNeil's) fore and aft lightwood tree, marked on the W side '50' very plainly." From that point, he set his quarter section post and proceeded to the western boundary of the Township. To make sure of the line, he ran six miles further west on the boundary of Township 2 North, Range 15 East, and back-tracked to double check his lines. As if this did not explain the total error, Hodgson remarked that the Whitten field was labeled a "rice field" in the McNeil

---

[5]Folder, "Reports of Draughtsman: Survey Work of Burr, *et. al*." Letter of November 22, 1850, Floyd to Putnam. Land Records and Title Section, Division of State Lands, Florida Department of Environmental Protection, Tallahassee, Florida.

notes. Conner informed the surveyor that Whitten's field was "in rice at the time" of the older survey. Whitten, who was with both McNeil and Hodgson, still lived within one half mile from this spot. Finally, Hodgson noted that in running the south boundary of the Township, he again found something peculiar, "In regard to the South Boundy. of same Township (tp 2 R 16) I traced same to the river from the E. crossed same, and after two hours Search in the pine woods, found it and traced it back to the River, and then discoverd the notch then reported in my Field Notes. How these notches could have occurred in McNeils work is more than I can account for, but that they are there, is beyound a doubt, and when the Boundy. line comes to be adjusted, between the States of Fla & Georgia, or Sooner, it will be found as per my work."[6] His prophecy would prove to be very accurate.

The notch or jog in the line persisted, even in Hodgson's revised traverse of the section in question. Floyd reported his findings of review to Putnam on March 24, 1851, "The notch, or jog, on the N. Bdy - (Georgia & Florida Line) still exists in the resurvey of Mr. Hodgson, and there is every evidence that it does really exist."[7] Putnam was now in somewhat of a quandary as to what to do. Upon receiving advice from Washington, he was allowed to hire another deputy surveyor to check on Hodgson's work. He chose one of the best, Arthur M. Randolph.

Randolph received his letter of instructions on February 2, 1852. He then notified Putnam that he would take the field during the first week of March, after finishing up some other engagements. As he stated, "I conceive my duty to be simply to examine the crossing of the Suwannee by the South Boundy. of T 2 R 16 N & E & the Georgia line in same T & R to ascertain if the jog reported by Mr. Hodgson certainly exists. I am not required to trace back & find <u>where</u> the mistake (if there is one) originated."[8] True to his word, Randolph began his examination in the first week of March 1852.

Randolph reported back to Putnam on March 25, 1852, that he had immediately found interesting and possibly confusing marks on trees. "There are two marked lines running parallel to each other & less than 20 chains apart. The lower or Southern is marked with blazes only, placed higher on the trees & not so frequent as on a well marked Surveyors line there is no evidence of its ever being

---

[6]Letters and Reports to Surveyor General, Volume 2, 1848-56. 945-47. Letter of December 16, 1850. Hodgson to Putnam.

[7]"Reports of Draughtsman: Survey Work of Burr, et. al." Letter of March 24, 1851. Floyd to Putnam.

[8]*Letters and Reports to Surveyor General*, Volume 2, 1848-56, 749-50.

posted or connected with the Township work & had its course not been that of the State line & its extension across the Suwanee river continuous it could scarcely have been mistaken for its more important neighbour. It is I presume the trial or random line first run by McNeil." Randolph then attempted to find the original marks of the McNeil survey, but noted that they were difficult to find. He also noted that but those marks that still existed were very clear and distinct. He observed that the lines running west from the river were fairly indistinct beyond the first few hundred yards and that only those of Mr. Hodgson appeared in the woods. From this Randolph concluded, "The absence of the slightest trace of a line in continuation thereof on the West side coupled with the fact that 9.87 chs North, there is a plainly marked line & a lightwood tree bearing figures refered to by McNeil to be conclusive evidence of a jog in the State boundary & correctness of Mr. Hodgsons survey." Randolph then went into a discourse on how the lines could be so different between two surveyors and noted that there was a discrepancy in the lines as run by Hodgson and C. C. Tracy, as reported by the Commissioner's letter of August 19, 1851. Here he defended his friend and observed that there was a slight jog in the first tier of townships north of the Basis Parallel, which resulted in other surveys being off somewhat. He then noted that McNeil's own statements concerning the lengths of the East and West boundaries of Township 2 North, Range 16 East. McNeil, he states, reported the length of the East boundary as 327.50 chains and the West as being 374.05 chains, "being an increase of 46.55 chs northing in running Six miles, or 7.76 chs (nearly) to a mile." As if to smile in irony, Randolph then observed, "It does appear rather strange that this should have escaped the zeal & vigilance of your predecessor in Office." As a final remark on the nature of the Hodgson and McNeil surveys of the line, he said, "The country is perhaps as bad as any in East Florida, wet, rough, saw palmetto woods cut up by, ponds, bays & swamps without number."[9]

The survey of R. W. B. Hodgson found the jog or error in the McNeil line that formed the boundary between Florida and Georgia. Randolph's examination verified the existence of the jog and because the jog was there a resurvey was needed to resolve the dispute between Georgia and Florida. Although Whitner's first associate did not agree with the colonel about the placement of the line, his second partner, Dr. Orr, did find grounds that were satisfactory to all parties.

The work of R. W. B. Hodgson was an important chapter in the story of the resolution of the Florida-Georgia line, one that has previously been unreported.

---

[9]*Letters and Reports to Surveyor General*, Volume 2, 1848-56, 753-59.

CHAPTER 12

CHARLES H. GOLDSBOROUGH

The process of surveying on an unknown frontier is difficult at best. The obstacles to be overcome are only to be guessed, the real or imagined "enemies" are sheer speculation and the securing of competent assistance is a variable to which there are no sure answers. In the frontier situation, these things can either mean disaster or fortune. To find your way into the great wilderness without some advanced knowledge of what lays ahead takes courage, skill and some luck. This is what faced almost every surveyor of the Florida frontier in the Territorial Period and later.

Not every surveyor or survey was successful. Some went broke and left the territory for greener pastures. Others tried their luck in the courts, and frequently lost! A few simply disappear from the written pages of history without much of trace. These frontier tragedies were often played out, to the extent of a written record, in the correspondence of the Office of the Surveyor General of Florida. The reading of these old letters gives a valuable insight to the life of the frontier surveyor and the problems that faced them in attempting to bring order out of chaos. One such failed attempt was that of Charles H. Goldsborough whose three-year struggle to survey the outer boundary of the Forbes Purchase ended in debt and humiliation.

The story of the Forbes Purchase is relatively well known to many in North Florida, however, a brief recapitulation is in order to help in understanding the nature of Goldsborough's problems. In essence, the "Purchase" was little more than a collection of debts from various Indian groups that owed money to the firm of Panton and Leslie, the greatest of the trading companies operating in the South during the British and Second Spanish Periods of Florida's history (1776-1821). The trading partners had worked with the Indians (Creeks and Seminoles) for many years and had accumulated large amounts of credit from these tribes in exchange for goods sold at their various stores, especially those on the Apalachicola River. With the consent of the Spanish governor, the firm was allowed to receive tentative title to the lands recognized as belonging to these Indians in return for the cancellation of the debts. The company did not have full right to sell any of these lands without the consent of the Spanish authorities. The bulk of the transaction regarding the land took place in 1804.[1]

The Panton, Leslie and Company changed its name to John Forbes and Company shortly after the cession of lands had been made in 1804. To it went all of the rights and privileges that had formerly been granted to the Panton, Leslie and Company. As Indian debts continued to mount in those tenuous years, the company had to petition to the Spanish governor for additional lands in return for the further cancellation of debts. By 1811, most of the lands to be included in the final grant were ceded to the firm, including the famous grant of Forbes Island, seven miles in length and one in width. Title was confirmed to the company by the Spanish government toward the end of 1811. This grant, or "purchase", encompassed nearly one and one-half million acres of land stretching from the Apalachicola River to the St. Marks River and as far north as Little River in modern Gadsden County.[2]

The grant actually implied that the Forbes Island was included and Richard Keith Call argued against this inclusion of the island when the case was heard before the Supreme Court in 1835. Call's arguments against the grant being confirmed to the Forbe's interests included the concept that the Indians did not own the land they gave in return for the dismissal of their debts (they belonged to the Crown), that the Spanish Governor of West Florida did not have the power to grant such lands and that the lands were in, technically, East Florida and out of the jurisdiction of the Governor of West Florida. An additional argument could have been made, as it was in the case of the second Forbes Grant), that the documents concerning the grant were fraudulent and therefore illegal. This was not forcibly argued in front of the Supreme Court and the documents produced by two visits to Havana were not admitted as evidence against the grant. When the ruling came down from Chief Justice Marshall, the island was included in the grant.[3]

---

[1]John C. Upchurch, "Some Aspects of Early Exploration, Settlement, and Economic Development Within the Forbes Purchase," (Unpublished Masters Thesis, Florida State University, 1965), 10-11.

[2]*Ibid*, 10-14.

[3]See William S. Coker and Thomas D. Watson, *Indian Traders of the Southeastern Spanish Borderlands: Panton, Leslie & Company and John Forbes & Company, 1783-1847* (Gainesville: University Presses of Florida), 1986. 350-360.

The second Forbes Grant was to extend to the west beyond the St. Andrews Bay and include nearly as much land as the first grant. The Innerarity brothers, James and John, had fought in local court to have it confirmed on behalf of the investors in the Forbes Company, of which they were two of the largest. When Call made two trips to Havana, he found his suspicions confirmed of the fraudulent nature of the second grant and brought the evidence in front of the court, which ruled against the validity of the grant, especially since it had been made after the deadline called for in the Adams-Onis Treaty of January 24, 1818. The second grant was, therefore, annulled by the court's action.[4]

The litigation over the grant was finally decided in 1835, in the case of *Mitchel v. United States*. The original suit had been brought in 1828 by Colin Mitchel and others in the Superior Court of Middle Florida on behalf of Mitchel's firm, a Savannah based land investment house which included John Carnochan, James Inerarity, William Calder, Benjamin W. Rogers and others. The firm lost the suit in the lower court and appealed to the Supreme Court of the United States, which has jurisdiction in cases involving treaties, etc. The boundary of the grant was defined by the Court's final order and included the Forbes Island portion and stipulated that if the attorneys for the claimants could show evidence that the Fort of St. Marks was included in the terms of the grant, they might include this property too. However, if the fort were used for military purposes and not part of the Indian cession to the firm of Panton, Leslie & Company, the fortress area would be public land of the United States and not part of the grant. So ruled favor the Superior Court of Middle Florida in 1838, a judgment that was affirmed by the Supreme Court of Florida in 1841.[5]

The Supreme Court's ruling in 1835 necessitated a survey of the boundary of the grant. Although the language of the Court's order made specific designations concerning the boundary line, it did not have an adequate map of the area nor did it have an official survey upon which to base any definite opinion. Therefore, the Surveyor General of Florida, Robert Butler, was ordered by the General Land Office to hire a surveyor to make the required survey so as to delineate the final boundary of the grant. Butler turned to Charles H. Goldsborough, a relative of Lieutenant Louis M. Goldsborough, the son-in-law to William Wirt, a former

---

[4]*Ibid*

[5]Apalachicola Land & Development Co. *et. al.* v. McCrea, Commissioner of Agriculture, *et. al.* 98 *Southern Reporter*. Cited in "A Report on the Application of the Marketable Record Title Act to the Sovereign Lands of Florida." Florida Department of Natural Resources, Office of the General Counsel, Tallahassee, Florida, August 1985.

Cabinet officer under Andrew Jackson. Goldsborough signed his contract to survey the grant on September 28, 1835, and was joined by Louis Goldsborough and Judge Richard C. Allen. The contract specifically read, "Connecting the Public lands with Surveying the several Deeds of Cession enumerated and recognized in the decree of the Supreme Court of the United States at the January Term 1835 To Colin Mitchel and others, as per copy thereof, and instructions furnished herewith." In layman's terms, this meant that all boundary lines surveyed under the grant's language must be tied into those already completed for the Public lands in the area and corners established where these lines met.[6]

The major direction given to Goldsborough was to follow strictly the dictate of the court and certified copies of the decision were given to him as part of his instructions. Butler also gave special attention to the surveying of Forbes Island, noting that, "The Island granted to John Forbes in the Apalachicola River must be also surveyed, and the necessary observation taken to show its connexion with same (the nearest stations) of the adjoining survey, with a view to perpetuate on paper in your return thereof its relative position in said River." He then continued, "you will designate in like manner St. Vincent or Deer Island, with the tract to which it is appended at the Mouth of Apalachicola River." Each grant, and there were numerous concessions mentioned in the Court's decision, was to be surveyed separately. This clause made the final tying together of all the surveys difficult, especially in consideration of the large number of islands, coves, rivers, streams, etc. that dotted the landscape to be included in the grant's survey. These special instructions also added one further complication, the boundary of the Fort of St. Marks could not be laid off until the Superior Court had ruled on its extent, but its exterior lines were to be run, "leaving the reservation to be bounded, where its limits shall be settled by the Court aforesaid." This unclear language made the eastern section of the survey even more confusing.[7]

The land greeting the surveyor was not the Elysian Field of classical mythology. Instead, it more closely resembled the name for a large section within the grant, later called Tate's Hell. The number of islands, the twisting, turning shoreline and the shear impenetrable nature of some of the lands made the attempt to run any meaningful lines difficult at best. On November 22, 1835, at the begin-

---

[6]Copy of Contract between Robert Butler and Charles H. Goldsborough, Dated September 28, 1835. Drawer: "U. S. Deputy Survyeors A-H," [File] "U. S. Deputy Surveyor, Charles H. Goldsborough." Land Records and Title Section, Division of State Lands, Florida Department of Environmental Protection, Tallahassee, Florida. Hereafter "Goldsborough file."

[7]*Ibid*, Letter of September 28, 1835. Butler to Goldsborough.

ning of his work, Goldsborough wrote to Butler asking for some specific instructions in surveying some of the questionable marsh near St. Marks:

> I do myself the honor to enquire of your Department whether you will require the Islands lying West of, and near the St Marks river meandered or not. These a half dozen very inferior group of Islands near that river, very boggy and covered entirely with marsh grass, indeed I question whether they ought to be considered Islands or as an appendage to the adjoining main land. I find on examining the map, which I presume the claimants had drawn, that the Islands which I have reference to, are not noticed, but considered simply as marsh land and attached to the main. There many little creeks or bayous which run in different directions and when during high water separate these Islands but when the tide is full ebb leaves but a mass of soft mud.[8]

This description of the land at the beginning of the survey was to presage the rest of this nearly impossible adventure.

Butler's immediate answer to Goldsborough's request for further directions was brusque and pointed, "Your letter of enquiry bearing the date of 22 Ulto. can only be answered by reference to the documents in your possession...From the description of the mud banks alluded to by you it would appear to my mind that they can not be esteemed other than marsh."[9] The tenor of the reply gives an indication that Butler was not going to be sympathetic to any delays caused by the surveyor not knowing a marsh from an island or any other reason.

By the following May, Charles Goldsborough's troubles were only just beginning. Not only did he incur the displeasure of the Surveyor General, but he had also aroused the ire of the holders of the grant, who, in turn, put more pressure on the over-wrought Butler. On May 7, 1836, Goldsborough penned the following letter in hopes of gaining more time, and, we suspect, sympathy from Butler:

> Sir: I have the honor to inform you of the very great difficulty I am at present labouring under, viz. that of employing men to assist me in the performance of the duties which you have been pleased to honor me with. I have had occasion recently to discharge three of my men, who, in consequence of disabilities I was <u>forced</u> to part with. I have to request you to permit me to suspend the survey, on which I am engaged,

---

[8]Goldsborough file, Letter of November 22, 1835. Goldsborough to Butler.
[9]Goldsborough file, Letter of November 30, 1835. Butler to Goldsborough.

until the ensuing fall, at which time I shall, as a matter of course, proceed to duty. I can at present, employ men at $40 per month which price I am not disposed to give, indeed I might say, unable to afford, and which I am satisfied you are well aware this contract will not authorise, taking into consideration the circumstances of my having been forced heretofore to give extra wages. However Sir; permit me to remark, that should you insist on the immediate execution of the duties on which I am engaged, I shall go with, notwithstanding the great difficulties I am laboring. With sentiments of great respect.[10]

Three days later, on May 10th, Butler fired off his reply:

Sir: Your letter of the 7 inst. is before me and I hasten to communicate that I would not feel myself authorized to suspend the execution of your contract until the fall ensuing upon any plea of pecuniary consideration and I feel it necessary to state that I had reported some time since to the Comms. of Genl. Land Office that I expected that contract completed in a short time. The claimants urge its completion the contract requires its completion, the Government expects its completion and my duties require me to say that I expected its completion before this time. Under all the circumstances I must beg your unremitted attention to this duty under apprehensions that you and securities will be ordered to be sued on your bond if much further delay shall be experienced.[11]

Butler's impatience with the constant delays in completing the contract are justified when it is considered that this survey was only expected to last one surveying season and not two years. Also, from the above letter it can be seen that the Surveyor General was under considerable political pressure to get the lands surveyed so the firm could commence sales of the land recently confirmed to it by the Supreme Court of the United States. Dispite the pressure asserted by Butler, the survey did not go on as scheduled.

The next letter in the correspondence between Butler and Goldsborough comes on January 21, 1837. Again we find Butler imploring the surveyor to report progress of any sort, "Sir: From the repeated solicitations and inquiries of the Attorney for the Company of the large grant Forbes & co and for the survey of which you long since entered into contract, I feel it my imperious duty to require that you will without delay report in writing the cause or causes which have led to

---

[10]Goldsborough file, Letter of May 7, 1836. Goldsborough to Butler.
[11]Goldsborough file, Letter of May 10, 1836. Butler to Goldsborough.

the unusual delay in filling said contract in all its provisions, that I may be enabled to report immediately a copy to the Commr. of the Gen'l L' Office for the information and decision of government."[12] Goldsborough replied on February 6:

> Your letter to me bearing the date 21st ulto. has just reached me, and in answer have to remark that the very great difficulty attending the survey of "Forbes Purchase" particularly that part which is required by your Dept. of making out exact returns showing the very many indentations on all the lines, also the exact area of all the purchases (say four) can be the only excuse I have in my power to offer you for the long delay of hand in the returns. I have been at work night and day for some time past at my brother's making the drawings of the survey, is the cause why your letter did not sooner reach me, and receive that attention which was incumbent on me on [Sic] (to) bestow on it. / You have been pleased, Sir; to remark that, "the unusual delay in filling said contract in all its provisions" &c &c. In answer to which I have only to say that the difficulty in procurring the necessary assistants and also the long spell of sickness which I was afflicted with last summer will possibly be a sufficient apology for the delay. It is only the person who performs the d___ery [drudgery] who knows and who can in all possibly appreciate the difficulty which I have had to encounter in the late survey. I was not aware myself, Sir, that more and serious difficulties awaited me at my table, by more difficulties have been, in a great measure, overcome, and I think I may safely say that you will be in possession of all the papers incident to the survey in the course of the present month.[13]

Two pieces of information stand out as reasons for the long delay in the completion of the survey, Goldsborough's personal illness over the summer and the lack of an adequate labor force on the frontier. He also notes other difficulties, but, at this stage, does not inform the Surveyor General exactly what they are. He also makes an unfortunate remark on Butler's knowledge of surveying on the frontier in stating that only the person who "performs the d___ery [drudgery]" can appreciate the difficulties. Butler, who knew the area as an aide to Jackson in his

---

[12]*Letters of Surveyor General*, Volume 2, 1836-41, 23. Bound volume of original letters. Letter of January 21, 1837. Butler to Goldsborough. Land Records and Title Section, Division of State Lands, Florida Department of Environmental Protection. Tallahassee, Florida. Hereafter, Letters of Surveyor General.

[13]Goldsborough file, Letter of February 6, 1837. Goldsborough to Butler.

Florida campaign and crossed some of the Forbes Purchase territory, was not one to take such an attempted slight lightly.

Butler, a good bureaucrat and administrator, did not want this delayed fulfillment of a contract to reflect poorly on him. On February 9, 1837, he wrote the Commissioner of the General Land Office, James Whitcomb, noting Goldsborough's letter, which he enclosed, and all of the back correspondence, plus instructions, to show that he was not the cause of the delay in fulfilling the Supreme Court's mandate.[14] On March 31, 1837, Butler again wrote to the Commissioner to note that he has had "to order Mr. Goldsborough again to the woods for the important purpose of connecting the public lands with the Survey of Forbe's et. al. as is required in his instructions...which he failed to comply with and which fact but recently came to my knowledge."[15] Thus, it appears from this letter that Goldsborough did send in his field notes for the survey along with the drawings, however, as Butler bluntly remarks, he did not fulfill the contract as instructed, therefore, Goldsborough was again sent to the field.

The costs of the survey were originally set at about four thousand dollars, according to Goldsborough's contract, however, Butler, evidently, did not think that sum would be expended. On April 20, 1837, he noted to the Commissioner: "It has been estimated that the Survey of Forbes Claim (now under execution) will cost near four thousand dollars, if this be true, a further sum than that on hand in deposit will be required."[16] Butler appears to have expected the survey to be done for less than the contracted amount, but, because of the delays, the added labor costs, return to the field, etc., Goldsborough had succeeded in running the costs close to this amount. The Commissioner wrote to Butler on June 7 acknowledging the receipt of his letter informing the General Land Office of Goldsborough's successful completion of the Forbes Survey and remitting $4,000 to Butler's account for payment on the contract.[17] On June 12, 1837, the Surveyor General again wrote to the Commissioner exclaiming that he now had enough money on hand to meet, "the expected demand of Mr. Goldsborough on his contract."[18] All

---

[14]*Letters of Surveyor General*, Volume 2. Letter of February 9, 1837. Butler to Whitcomb. 23.

[15]*Letters of Surveyor General*, Volume 2. Letter of March 31, 1837. Butler to Whitcomb. 26.

[16]*Letters of Surveyor General,* Volume 2. Letter of April 20, 1837. Butler to Whitcomb. 29. Also see the contract in Goldsborough file.

[17]Letters from Commissioner, Volume 2, 1832-39. Letter of June 7, 1837. Whitcomb to Butler. 419. Land Records and Title Section, Division of State Lands, Florida Department of Environmental Protection. Tallahassee, Florida.

[18]Letters of the Surveyor General, Volume 2. Letter of June 12, 1837. Butler to Whitcomb. 36.

appeared to be finished with the long awaited survey of the Forbes Purchase, but it was not to be.

From the correspondence of the Surveyor General, dated June 29, 1837, Goldsborough apparently had not turned in the final product of his survey. Butler expected the results almost daily and was anxious to have done with this problem.[19] On July 3rd, Butler again wrote to the Commissioner informing him that he had met with Goldsborough and was informed by the surveyor of, "being ready to return his work into the office in a week at farthest."[20] Seven days letter Butler received the following from the tardy surveyor:

> I have the honor to acknowledge the receipt of your letter of this morning calling on me for a circumstantial statement showing the condition of my contract. It is a subject Sir; of regret to me that I should have been so long beyond the period for the completion of my contract. In the protraction of the extensive grant to Panton, Leslie & Co. including all the lands between the Apalachicola River on the West and extending to the River St. Marks on the East I find on examination that there has been a great error, so much so, that I could not, with propriety had in your office. The plot has been protracted and as I thought a day or two ago, would be ready for examination by this time, yet it requires a more thorough one than I have enabled to bestow on it and also make out a new plot entirely. The one I have finished being drawn in ink it will require at least a fortnight to protract another so as to ascertain where the error was committed [t]here being upwards to <u>three thousand</u> courses. All the field notes are finished and ready for inspection. The plots of the other grants are also finished and ready for inspection....I will proceed immediately to the examination and protraction of the above named grant and when finished will report to your office the result of it.[21]

The delay produced by this "great error" was to be costly. Two days later Goldsborough again wrote Butler, this time with a new request.

The letter of July 12th shows the extent of expenses incurred by the Deputy Surveyor in attempting to finish this enormous survey. The letter and Butler's response also indicate the closeness usually found between Butler and his deputies:

---

[19]Letters of Surveyor General, Volume 2. Letter of June 29, 1837. Butler to Whitcomb. 40-41.
[20]*Ibid*, 41.
[21]Goldsborough file, Letter of July 10, 1837. Goldsborough to Butler.

> Owing to the great length of time I have been engaged in my contract with your office and the very heavy expense I have of necessity, been at, which are upwards of $2500 induce me again to ask you for such an advance of which you are to be the judge. I feel rather diffident in pressing the matter on you, but Sir; I must resort to the old adage that there is not excuse necessary for an application of this character when necessity demands it of which, I assure you is fully the case. I pledge myself that the returns of my contract will be forthcoming.[22]

Butler replied on the same day:

> Sir: Your letter of present date is just received requesting an advance of public funds on your contract for the survey of the claim decreed by the Supreme Court to certain individuals therein named. Will you have the goodness to revert to the conversation had with you on this subject a few weeks since wherein I informed you most distinctly that my instructions forbid me in positive terms to making such advances. If I had private funds to spare I would take pleasure to relieve your wants (as I have before to others without the advantage of one cent emolument) but I am denied the pleasure of obliging you.[23]

At the same time, Butler wrote to the Commissioner of the General Land Office that Goldsborough had found the "great error" in his own work and that it would, "require some time to correct."[24]

By mid-October the returns of Goldsborough's survey had not been sent to Butler and he had to inform the Commissioner that such was the case. He also noted that should the returns be sent in soon, they would still have to await the decision of the Superior Court of Middle Florida concerning the boundaries of the St. Marks reserve, which would then have to be surveyed and included in the Goldsborough contract.[25] This an even longer time before the money-starved Apalachicola Land Company, the successor to John Forbes & Co., could get their lands to market with a possibility of clear title.

With time and patience running out, Butler received the following from Goldsborough on November 28th:

---

[22]Goldsborough file, Letter of July 12, 1837. Goldsborough to Butler.

[23]*Letters of Surveyor General*, Volume 2. Letter of July 12, 1837. Butler to Goldsborough, 45.

[24]*Letters of Surveyor General*, Volume 2. Letter of July 12, 1837. Butler to Whitcomb. 46.

[25]*Letters of Surveyor General*, Volume 2. Letter of October 16, 1837. Butler to Whitcomb. 47.

> In case you may suppose me too negligent (in which, 'tis tru, I have been) in the performance of the duties prescribed to me by your office in the year 1835, I have the honor to make you acquainted in detail with the circumstances which produced the failure of making out the returns which are usual with your deputies. In surveying the Appalachicola River I was forced as a matter of necessity, to observe the triangular mode of survey, and in doing so, I must have made a considerable error & am therefore unable to form the plots or rather close the maps in accordance with the usage of your Department. I have, <u>in vain</u> tried every means to close the maps so as to make a correct survey, but find it utterly impossible. A re-survey will be absolutely necessary & I should have undertaken it myself long since had my health have permitted it, which has been bad, and is at this time in rather a precarious situation, and I have my doubts whether I could survive the swamps of Appalachicola were I now to go in them. The other surveys/the one West of the River is completed also the Island the Appalachicola and the Islands appertaining to the large survey together with notes of those surveys.[26]

This letter led Butler to ask the Commissioner to decide whether he could hire another surveyor to finish the job at Government expense or take it from the contracted amount due to Goldsborough through a suit for non-compliance.[27] One final complication arose from the reports of murders of settlers on the Gadsden and Wakulla frontiers. This meant that no surveyors would be able to take the field and complete any survey at that time.[28]

Not until February 1, 1839, did Robert Butler get the opportunity to hire a deputy to finish and correct the work attempted, but never completed, by Charles Goldsborough. On that day, he hired Robert Ker, a fellow member of the First Presbyterian Church of Tallahassee and experienced surveyor. The contract reads almost exactly as that of Goldsborough's, as does the special instructions inclosed with the contract.[29] As Rod Maddox, Public Land Surveyor, noted a few years ago in his, "The Forbes Purchase: A Surveyor's Dilemma, "First, the entire boundary was run as a tremendous closed traverse, and the three other contiguous grants given in 1811 were run also along with the islands off the coast which were given in the first cession. This survey [Ker's] is the first complete boundary sur-

---

[26]Goldsborough file, Letter of November 28, 1837. Goldsborough to Butler.
[27]*Letters of Surveyor General*, Volume 2. Letter of November 28, 1837. Butler to Whitcomb. 53.
[28]Letters of Surveyor General, Volume 2. Letter of October 1, 1838. Butler to Whitcomb. 121.
[29]Contract of February 1, 1839. Robert Ker. " Drawer: U. S. Deputy Surveyors: I-N," "File: U. S. Deputy Surveyor.

vey whos notes are presently on record. He then [Ker] retraced the original section, township and range lines which the government did complete on the exterior of the Purchase boundary, and tied them together with the boundary."[30] These tasks, not performed well by Goldsborough, were satisfactorily completed by Ker during the 1839 surveying season. With the survey complete and acceptable, Robert Butler could rest more easily and turn his attentions to the many other important surveys to be completed under his direction.

---

[30]Rod Maddox, "The Forbes Purchase: A Surveyor's Dilemma," 11. Paper on file in the Land Records and Title Section, Division of State Lands, Florida Department of Environmental Protection. Tallahassee, Florida. This paper was originally composed as a requirement for seniors in the surveying seminars in the University of Florida's surveying program and was done under the supervision of Professor David Gibson.

CHAPTER 13

FORGING THE FLORIDA FRONTIER: THE LIFE AND CAREER OF CAPTAIN SAMUEL E. HOPE

The frontier is often defined as the area beyond the edge of civilization or settlement, the edge of the wilderness or some similar notation. It is much more than these definitions in that it also shapes the character of those who try to tame it. The frontier is also an area to be exploited and developed by those who claim it as their dominion. Fortunes and lives can be made or lost in the struggle the frontier demands before it gives up its riches. Life on the frontier is harsh, sometimes barren, often lonely and frequently brutal. The only sugarcoating offered by the frontier comes with the cane the pioneer plants, nurtures, cultivates and grinds himself. Yet, by the acts of planting, nurturing, cultivating and refining the frontiersman brings forth the civilization and settlement, now redefined by the new circumstances, similar to that the frontiersman once left behind.

The biggest change in the circumstances of the new settlement and civilization is the frontiersmen who created it are now the new leadership. Men who once were counted on to follow or bow to the established order are now those making the rules that they, in turn, expect others to obey. The constant flux of the frontier situation brings with it more conflict between those who wish to stake their claim to the roles of leadership. However, because the new settlement is on the edge of the new frontier, it is still open to the violence and danger lurking just beyond the line of sight. The question facing all such settlements is can the new leadership actually take the next step and create a more stable and less violent community? It is into such a situation that the father of Samuel E. Hope, William Hope, stepped in around the year 1842.[1]

William Hope was a man of great determination and stamina. Born in Liberty County, Georgia in 1810, William moved his family to Florida in the mid-1830s. Accounts differ as to the first settlement of William Hope, however, he was soon engulfed, as was all of Florida, in the conflict known as the Second Seminole War.[2] After service in the Florida Militia and the drawing to an end of this tragic conflict, he moved his family to the area near Brooksville, on the edge of the Choocachattee Hammock at a place to be named, "Hope Hill."[3] Richard J. Stanaback, in his *A History of Hernando County, 1840-1976*, noted that Hope became a substantial rancher and regularly drove his cattle to the Tampa market where he sold his beef, "for a handsome profit." Citing the *Florida Census for 1850*, Stanaback quotes this work as detailing Hope's family and holdings. His family included his wife, Mary Jane, the second Mrs. William Hope, 22; Samuel, 17 years of age and a "student" and born in Georgia; Virginia, 10 years and born in Florida; Adela, 3 years of age, also born in Florida; and, finally, baby, listed as a female, Christiana. By 1850, William Hope was listed as a planter who owned

---

[1]"Biographical Sketch of Samuel E. Hope," Clara Hope Baggett, Printed in 1919. A copy can be seen in the Hope Family file at the Hillsborough County Historical Commission, Library and Museum, Tampa, Florida. A copy was provided to the author by Kyle Van Landingham, to whom the author is deeply indebted. Mr. and Mrs. L. E. Vinson, of Tarpon Springs, descendants of Captain Samuel Hope, have also provided this sketch, along with countless other documents which are used in this short biography. Without the aid of the Vinsons, this article would not be possible. Donald Ivey, Curator of the Heritage Park Museum in Largo, Florida, has also provided much useful and informative data from the files of the museum. And, finally, the generous staff and fine collection found at the Tarpon Springs Area Historical Society, was of great benefit to this article.

[2]A genealogical chart drawn up on Samuel Edward Hope in the Heritage Park, Pinellas County Historical Museum, states that William moved to Jacksonville, while Clara Hope Baggett's sketch states they went to Brandy Branch, northwest of Jacksonville. Both sources, when read carefully, agree that William Hope migrated to Florida in 1836.

[3]William Hope appears on the muster roll of Captain J. G. Black's Company of Florida Mounted Volunteers. This group was mustered into service on May 9, 1839, and mustered out, in the usual six months time, on November 9, 1839, at Fort Harllee. He served as a private. Also appearing on the roll are John C. Hope, 1st. Lieutenant and privates William Hope Jr. and David Hope. Florida Department of Military Affairs, Special Archives Publication Number 68: "Volume 2, Florida Militia Muster Rolls Seminole Indian Wars," State Arsenal, St. Francis Barracks, St. Augustine, Florida. 34-35. [No date of publication]

2240 acres of land, a substantial holding for a frontier farm.[4] According to Clara Hope Baggett, William Hope owned one hundred and fifty-seven slaves by the time of the Civil War.[5] Given the frontier nature of the area, there can be no doubt that William Hope's determination to succeed in his new environment paid off handsomely for the family. Also, his newly acquired status brought him respect from his neighbors and made his one of the "most influential families" in the area.[6]

Samuel Edward Hope, the only male child born to William Hope and his first wife, Susan Mitchell Harville, was born on September 17, 1833. Only three years of age when William moved to Florida, the trip to Brandy Branch, near Jacksonville, was probably very exciting. The Second Seminole War brought hardships to nearly every Floridian, Indian and white alike. Food shortages were common and the federal government instituted a program to feed those who suffered at the hands of Indian attacks. It is difficult to believe that Sam Hope had the opportunity to carry on "normal" studies to prepare himself for the next recorded step of his life, study at the Alexandria Boarding School in Alexandria, Virginia. At the school, Sam Hope did very well in Plane Trigonometry and Surveying and Davis' Algebra, but did not fare so well in reading, grammar and spelling for the term ending July 1, 1854. His overall marks were, however, high enough to earn him his certificate of proficiency, an equivalent to the diploma by today's standards.[7] The fact that his highest scores were in the field of surveying gives an indication of his future direction and interests.

Sam Hope returned to Florida to find it again preparing for war with the remaining Indian population. By 1856, the war had begun in earnest and Sam Hope was called upon to serve. His first appointment was none too glorious, that of First Lieutenant on the staff of General Jesse Carter and assigned as "Special

---

[4]Richard A. Stanaback, *A History of Hernando County, 1840-1976* (Brooksville: Action '76 Steering Committee, 1976), 13-19. Stanaback states that Hope's first settled in Florida south of Gainesville, but later moved near Newnansville. This seems highly unlikely given that the former city did not exist when Hope entered Florida and "south of Gainesville" would have meant Micanopy or Wacahoota. Neither of these two settlements list Hope as an early settler.

[5]Clara Hope Baggett, 1.

[6]Applications for Employment, Volume 3: 1858-1860. Letter of June 22, 1858. H. V. Snell to F. L. Dancy. 79. Florida Department of Environmental Protection, Division of State Lands, Land Records and Title Section, Tallahassee, Florida.

[7]Copy of Samuel E. Hope's Certificate from the Alexander Boarding School, Thirtieth Annual Session - Ending 7th mo. 1st 1854. Copy in the possession of the L. E. Vinson family and used with their permission.

Agent to the Independent Companies of Volunteers." His specific assignment was as Assistant Quartermaster and Coroner for all frontier troops. This duty lasted from October 22, 1856, to February 22, 1857.[8] His second tour of duty brought him into contact with one of the more energetic men on the South Florida frontier, Hamlin V. Snell. Hope served as First Lieutenant under Snell's company of Mounted Volunteers from December 15, 1857, until this unit was mustered out on May 22, 1858, at the end of active hostilities. The muster roll for this unit shows young Hope to be twenty-four years of age with a horse worth $200 and equipment worth $15, a relatively expensive outfit for time and vicinity.[9] Snell recognized the character of his former First Lieutenant and when Sam Hope applied for his first job as a U. S. Deputy Surveyor, Hamlin V. Snell wrote to Surveyor General F. L. Dancy the following, "I take this method to commend to your favourable consideration Mr. Samuel E. Hope of Hernando County who wishes an appointment as Deputy Surveyor. Mr. Hope is a staunch Democrat and connected with the most influential families of his County his qualifications are of the best order and his habits are unexceptionable his appointment will confer a favour upon your obt servant."[10] Dancy, a West Point trained engineer, staunch Democrat and fellow officer in the Florida Militia, in which he held the rank of Colonel, understood the importance of appointing someone with Sam Hope's connections.

On December 27, 1858, Sam Hope began the professional career that was to last through most of his active life in one form or another. With the receipt of a letter from Dancy dated December 16, he accepted the obligations for the contract he soon signed and presented in person in early January 1859. At that point in his life, Sam Hope had never done a survey on his own, however, he wrote to Dancy, "I have never had much practice in surveying, but I have studied the theory and feel confident that I can give satisfaction as to my work."[11] Sam Hope received his instructions on January 15, 1859, and headed south to his appointed region in

---

[8]Florida Department of Military Affairs, Special Archives Publication Number 67. "Florida Militia Muster Rolls, Seminole Indian Wars, Volume 1," 1.

[9]Florida Department of Military Affairs, Special Archives Publication Number 75. "Florida Militia Muster Rolls, Seminole Indian Wars, Volume 9," 96-99.

[10]Applications for Employment, Volume 3: 1858-1860. Letter of June 22, 1858. H. V. Snell to F. L. Dancy. 79. Florida Department of Environmental Protection, Division of State Lands, Land Records and Title Section, Tallahassee, Florida.

[11]*Letters and Reports to Surveyor General*, Volume 3, 1857-1861. Letter of December 27, 1858. Hope to Dancy, 47. Florida Department of Environmental Protection, Division of State Lands, Land Records and Title Section, Tallahassee, Florida. Hereafter *Letters and Reports*, date & page number.

the area near the Kissimmee River and Lake Istokpoga. The land he surveyed was swampy, filled with creeks and ponds, and subject to inundation during the rainy season. On April 2, 1859, he wrote confidently to the surveyor general from Fort Meade, "I found the country very dry and I got along better than I anticipated. I have finished all in Tp 34 & 35 Range 30 & 31 S & E. I am now on My way down Peas Creek [River] in Tp 38 & 39 Range 24 & 25 where I am expecting to find a better country than I have been through." He also noted for Dancy's benefit, "Some think that the Indians will be apt to trouble me while on this work, but I intend to give them the trial of it. I wont quit until I am made, until I get through."[12] He had no worries about the Indians and saw none during his survey. Yet, confident though he may have been in early April, by May 16th he was reporting that he had returned home and had not completed the contract, because one township remained to be finished.[13]

Hope had a good reason for not finishing his survey and it was one that Dancy, as a leader of the State militia, would easily understand and forgive. "The cause of my not finishing was this," he wrote, "I was Liut. in Capt. Snells Co. Mt. Vol. and in making up the company I had to become responsible for several thousand dollars for horses in the company. I was ready to commence work on my last Tp when my Father sent for me, that the Paymaster was in Tampa and paying off the troops. I had then to quit to go there and my provisions being only enough to last me the time it would have taken me to finish and get to where I could get some—which was Tampa—I found in getting to Tampa that my mules would never hold out to get back to my work, and I concluded that I would fix up my work that I have and ask for time to finish the other, if nothing else but finishing it will give Satisfaction. I wish to give satisfaction to my work if I dont make any thing on my contract." Dancy, as expected, gave Hope an extension of time to finish this difficult contract. The four townships in the area of Lake Istokpoga were very difficult. As he explained, "I dont think that any one man ever had four Townships like the four I had on Lake Istokpoga. I dont think there ever could be that number together again or before so bad as they were." With his usual expression of dogged determination, Sam Hope continued, "I never like to take hold of any thing and fail. If I take a contract and make a hard bargain I always Stick the tightest to it."[14] Hope finished his survey but had one request when offered

---

[12]*Letters and Reports*, Letter of April 2, 1859. Hope to Dancy. 66.
[13]*Letters and Reports*, Letter of May 16, 1859. Hope to Dancy. 71-72.
[14]*Ibid*

another surveying contract the following season, "Can you give me the subdividing of Townships 37 & 38 or 39 S, Ranges 26 & 27, with the other Townships & Ranges given in the Bond leaving out the Ranges 29 & 30 in Township 36, <u>anywhere in the whole country but on Lake Istokpoga</u>."[15] The tough, wet sawgrass prairie of that region was simply too much, even for a tough frontiersman like Sam Hope.

Hope's inexperience in surveying was soon overcome by actual fieldwork and he became skilled enough to spot errors in his own work. "I have finished Township 39 S Range 24 E, Tp 40 Range 23, 24 & 25," he reported to Dancy on February 21, 1860. "I have found out the error in Tp 39 S R 25 East of my last year work, it was an error of my own and a large one at that, but I am more than willing to correct my error in my work. I though that I was correct but I was not. I am now at work on that Township and will give you new set of field notes for the whole Township when I get through."[16] This extra work was done at his own expense, which meant paying the crew's wages, the costs of the food, transportation and instruments and all other costs associated with drawing up the revised field notes and plats. The profession of surveying, even in the earliest years, had substantial up-front overhead costs which had to be borne by the surveyor. This meant that most of the early surveyors had to have some wealth to perform their contracts or be backed by those who did, most often indicated by those who backed the surveyor's personal bond. In Sam Hope's particular case, his bondsmen were William and David Hope and William Wall, more evidence of the success of William Hope in shaping the frontier situation into personal benefit.[17] The family characteristics of stamina and determination showed in Sam's attitude toward his surveying career and the recognition of his own error and the willingness to correct it.

The year 1860 brought great changes to the life of Samuel E. Hope. After finishing his second surveying season, he married Mary Henrietta Hooker, the daughter of William B. Hooker, one of the most prominent men in Florida. A woman of refinement and an above average musical talent, Mary Hope was educated at the Soventon Masonic School in Georgia. The fact that Sam Hope was also a member of the Masonic Lodge did not hurt his chances of winning Mary's

---

[15]*Letters and Reports*, Letter of November 3, 1859. Hope to Dancy. 99-100.

[16]*Letters and Reports*, Letter of February 21, 1860. Hope to Dancy. 112-13.

[17]Drawer, "U. S. Surveyors A-H," [File] "U. S. Deputy Surveyor Samuel E. Hope." Contracts and Bonds. Florida Department of Environmental Protection, Division of State Lands, Land Records and Title Section, Tallahassee, Florida.

father's approval of the match. This union produced eight children and many years of communal happiness for the Hope family.[18] However, the newlyweds nearly had a very short marriage when, on April 25, 1860, they were reported as injured in a train-wreck near Lake City, Florida. Sam dislocated his shoulder and Mary suffered foot lacerations, both, luckily, short lived injuries.[19]

This same year also saw Sam Hope enter the field of politics. The Tampa-based *Florida Peninsular* for September 1, 1860, published the first announcement of his entry into the race for the Hillsborough seat in the State Senate. His opponent was not, like Sam, a political novice but the highly experienced local politician, James T. Magbee. Because the Democratic party was highly fractionalized at this point in time, primarily by the brief but colorful Know-Nothing party, the field for office appeared to be wide open. Some of those wanting county division, a splitting off of Hillsborough and Manatee with the creation of a new county, later called Polk, backed Magbee, who was the first to announce his candidacy. Many opposed to division, such as Hope's old commander H. V. Snell and Madison Post, the former mayor of Tampa, both bitter enemies of Magbee, who had engineered his removal from the post of collector of customs with Snell as his replacement. Post, to use historian Canter Brown, Jr.'s phrase, "attempted to pull together and manipulate against Magbee the Hillsborough County Democratic party." The result, as Brown notes, was a shallow attempt at a party convention where Hope was nominated by Snell, the "convention" being poorly attended and many communities not represented.[20] The conservative *Florida Peninsular*, edited by former judge, Simon Turman, came out quickly on the side of Magbee and lashed out at Hope for following Post's leadership. Charges and counter charges of office bartering, hypocrisy and political manipulation literally covered the pages of the *Florida Peninsular*. Although some have viewed this as a relatively gentlemanly affair, and by frontier standards it may have been, it was still an old fashioned "mud-slinging" campaign by all accounts. With the help of Turman, John Darling and other leaders of Hillsborough's pro-division forces,

---

[18]Clara Hope Baggett, 2.

[19]*Florida Peninsular*, May 26, 1860, 3.

[20]Canter Brown, Jr., *Florida's Peace River Frontier* (Gainesville: University Presses of Florida, 1991), 131-34. Brown loosely demonstrates the volatile nature of Magbee's career with its many political swings, including a switch to the Republican Party after being, at first, a Union Democrat and changing to a staunch backer of the Confederacy, even serving in the 1865 Constitutional Convention along side of Sam Hope. Unlike Magbee, Hope, the "staunch Democrat," remained one until he died.

Magbee won by a substantial margin.[21] Hope's year had started in a prosperous fashion but came to an end in an unsuccessful election bid, but the fact that he had learned much about politics, gained a political base and married the woman who would become his lifelong companion had to ease the brief pain of the election loss.

Sam Hope did not have long to relive the lost election for other, far more important matters, were to enter into the lives of all Floridians. Florida, in December 1860, elected members to a convention that was to decide the fate of the State early in the following year. The decision to succeed from the Union was not easy and left much of Florida divided. Recent studies have indicated a great deal of Unionist sentiment at the time of the convention. Former leaders, like Territorial governors Richard K. Call and Thomas Brown openly opposed any break with the Union and local leaders, such as F. A. Hendry, Jacob Summerlin, Ossian Hart and James T. Magbee, were all supporters of keeping Florida's ties to the Union.[22] Although Hope may not have wanted war, he certainly supported the principles upon which the Confederacy was founded and did not shy away from his duty to defend his home state.

Sam Hope's Civil War duty extended from early in 1862 until the very end of the war. His first assignment was with the local unit named the Brooksville Guards which were organized locally on February 22, 1862, and was mustered into Confederate service on March 15, 1862, with Samuel Hope elected as its captain, for a term of "3 years or during the War."[23] The duty called for the unit to defend the area around Bayport, Florida. The duty, except for scouting the coast and peering at the ever-tightening Union blockade, was dull and tedious. The company was assigned to Bayport for about one year when it was transferred to Tampa for nearly one year. The only break in this dreary duty was a brief sojourn to Crystal River in August of 1863. In the spring of 1864, Sam Hope's company was ordered to Camp Finegan, about seven miles west of Jacksonville. Here,

---

[21]*Florida Peninsular*, September 8 through December 8, 1860. The issue for December 8, 1860, carried a number of important letters and editorial comments concerning this election, which had been decided in early October. The fact that the recriminations carried on for so long after the election gives a clue to the bitterness of the race.

[22]Brown, *Florida's Peace River Frontier*, 140.

[23]Letter of January 22, 1910. Sam Hope to Mrs. J. C. Davaut. "United Daughters of the Confederacy: Florida Division (Papers), Volume 1." Mrs. Townes R. H. Leigh, compiler, 1926-27. State Library of Florida, Florida Room (Dodd Room). Tallahassee, Florida. Hope's Confederate Record lists him as a Captain of Company C and the date of Entry or Muster into the Confederate service as March 15, 1862.

Hope's company came face to face with enemy troops for the first time.[24]

This affair, brief as it was, did not end gloriously for the 6th Florida Battalion, of which Hope's company was now a part. Although Sam's men were willing to fight the enemy as they approached the camp, they soon discovered that the Federal troops had taken two roads to reach the area and threatened to cut off other units of the 6th Battalion. With the threat of being surrounded by the enemy starring him in the face and because his unit was greatly outnumbered, Hope relied on a soldier, W. L. Eubanks, who was from the immediate area, to guide them to safety and tie up with other units, most notably John W. Pearson's Ocklawaha Rangers. After successfully evading the raiding federals, Hope's company and the remainder of the 6th Battalion headed for Lake City, arriving one day before Florida's largest battle was to commence.[25]

The Battle of Olustee, or Ocean Pond, took place on February 20, 1864, on a site not selected by either commander. The 6th Battalion was stationed on the extreme right of the battlefield near the railroad tracks. "There it opened a deadly enfilade on the 8th Colored Troops," states historian Gary Loderhose, "inflicting such severe damage as to compel them to fall back in mass confusion, abandoning five pieces of artillery in the process." Then, as so often happened in this war, the ammunition ran low and a halt in the fighting was called until supplies were brought forward. By this time, the battle had been, in essence, won and the Union invasion of Florida ended.[26] Sam Hope's Company C was in the thick of the

---

[24] Gary Loderhose, "A History of the 9th Florida Regiment," (Unpublished Master's Thesis, University of Richmond, May, 1988), 30-32. Loderhose does not mention any return to Tampa after the Crystal River service and has the unit immediately going to Camp Finegan. Hope states that his company was not transferred to the camp until spring of 1864. The records, thus, show an unaccounted for gap as to where the company, reduced to 36 men at Crystal River, was stationed between August-September 1863 and February 8, 1864, the date Hope states that they lost equipment in the evacuation of Camp Finegan. See Records of Samuel E. Hope, Company C, 6th Battalion Florida Volunteers. Military Records of the 9th Florida Regiment. Records of the U. S. War Department. Record Group 109, National Archives. Washington D. C. Copies of these records were supplied to the author by the generosity of Kyle VanLandingham.

[25] Zack C. Waters, "Florida's Confederate Guerillas: John W. Pearson and the Oklawaha Rangers," *Florida Historical Quarterly*, 70 (October 1991), 143-44. Mr. Waters has a typescript of the "Reminiscence of Captain Samuel E. Hope" in his personal possession, but has been generous in providing a copy to the Pinellas County Historical Museum at Heritage Park, Largo, Florida. Donald Ivey, curator of the museum, was kind enough to lend me a copy of Mr. Waters' typescript for this article.

[26] Loderhose, 53-55. The best account of the entire battle can be had by reading David J. Coles, "A Fight, a Licking, and a Footrace: The 1864 Florida Campaign," (Unpublished Master's Thesis, Florida State University, 1985).

fighting, suffering heavy casualties and recording the highest death rate of all the companies.²⁷ Hope, himself described the fight and aftermath as follows:

> I went on the Battlefield on the day of the fight with 30 men all told in my company. We were the Color Company of the Battalion and in that fight at Olustee lost 15 men Killed and wounded 5 killed dead and 10 wounded. The night after the Olustee fight I was ordered forward to St. Marys River to take care of a tressel after the Union Army had retreated to Jacksonville.²⁸

The total for the entire battle has been given as 1,861 killed, wounded or missing for the Union forces, while Confederate losses totaled 946. The over 2,800 casualties of this battle make it more than a minor skirmish, though it does pale in comparison to Gettysburg, Shiloh or many of the Virginia battles with which most people are familiar.²⁹

Sam Hope's time in Florida, effectively, came to an end with the Battle of Olustee; however, two events that took place during his Florida service have made him a minor legend in the history of his home state. The first incident took place, allegedly, in the late summer of 1864 when some Union deserters, along with some escaping slaves, attempted to send a signal to a passing blockading ship from the mouth of Anclote River. According to local historical writer, Glen Dill, "However, a tough Confederate captain was hot on their heels with a small detachment of soldiers. Waiting for low tide, they crossed the river at night, surprised the fugitives, and hanged them all on the spot."³⁰ According to Wilfred T. Neill, in an article published in the St. Petersburg *Times* on February 19, 1978, the deserters were "dissatisfied" Confederates who were attempting to flee service at Fort Brooke. Again, the date is given as mid-1864.³¹ To this day, the story of Sam Hope and the hanging of the deserters on Deserter's Hill persists. There is only one catch to the story, Sam Hope's only leave in 1864 came in February of

---

²⁷Loderhose, 55.
²⁸Letter of January 22, 1910. Hope to Davaut. UDC Papers.
²⁹Loderhose, 54-55.
³⁰Glen Dill, "Sorting Through the Stories Behind Deserter's Hill." Copy obtained from the clippings file at the Tarpon Springs Area Historical Society, Tarpon Springs, Florida. No date.
³¹Wilfred T. Neill, St. Petersburg *Times*, February 19, 1978. Also from the clippings file at the Tarpon Springs Area Historical Society, Tarpon Springs, Florida.

that year and lasted only twenty-eight days.[32] He did not return to the area until he was elected to the Legislature in October, therefore, he could not have been around when the accounts allege that he hanged the fleeing deserters. If the incident did happen, as local tradition insists, it had to have occurred during this February leave, not in "late summer" as Dill's account has it. Until further documents surface definitely linking Hope to the hanging of the deserters, there is no reason to believe that Hope was the commander of the unit that perpetrated the hangings.

The other incident involved Hope in the alleged disappearance of one Henry M. Stanley, the famed explorer of Africa, from the Confederate forces on the verge of the Battle of Olustee. According to an article published in Pensacola on May 31, 1913, Hope reportedly told reporter Frank Huffaker, "The last time I saw Stanley he was gathering his belongings preparatory to decamping, and I think he stayed behind just long enough to get captured and sent north." Hope told Huffaker that Stanley had joined his command at Tampa in late 1863 or early 1864 and that he was so, "awkward and English in his ways that the other boys dubbed him 'Darby Gallikins,' and that name stuck to him until he disappeared. Hope noted that he was a "rawboned fellow" with the makings of a good soldier, however he was either captured or deserted to Union forces when Seymour's troops captured Camp Finegan. When asked whether he was sure this was the same Henry M. Stanley of exploration fame, Hope replied, "Sure, do you think a fellow could ever forget that Englishman after looking at him once?" Sam Hope speculated that things were just too slow in Florida and that Stanley wanted to get up north to Virginia, where the fighting was hotter. M. N. Hill, another Anclote resident who served with Hope, was also interviewed about Stanley and declared he was a member of his "mess" during the campaign. He agreed with Hope's assessment and believed he submitted to capture so as to be sent north, where the real action was.[33] Whether true or not, the story made for entertaining reading in 1913 and still fascinates the curious today.

In early March 1864, Major General Patton Anderson, the new commander of the districts of East and West Florida, received a call for more troops to be sent to the Virginia theater of the war. Grant's strategy of wearing down the armies of

---

[32]Letter dated January 25, 1864. Leave was granted from February 1 to 28th, 1864. Record Division: Rebel Archives, War Department (stamped): Record Group 109, Military Records of the 9th Florida Regiment. National Archives: Washington, D. C.

[33]Pensacola *Morning News*, May 31, 1913. Copy from the clippings file at the Tarpon Springs Area Historical Society, Tarpon Springs, Florida. Name of the newspaper is unclear.

Lee was enjoying success and men were desperately needed at the major front. At this time, Sam Hope's men had become part of the newly formed 9th Florida Regiment of Infantry. On the May 18, the new regiment pulled up stakes and marched into Georgia to catch the trains that would take them to their new destination in Virginia. The trip was anything but plush and rations were shorter than the men's patience. After many stops and little food, the 9th Florida Regiment arrived at Petersburg, Virginia, on May 24, 1864.[34] The fortunes of the troops were to now take a decidedly different and deadly turn.

The new arrivals were immediately assigned to the division commanded by Major General Anderson, however, since he was absent, the command passed to General William Mahone of Virginia. In the first week of June, the 9th Regiment, now part of the unit called Finegan's Brigade, was involved in fighting Union forces under General Philip Henry Sheridan and Hope's company suffered two casualties, both of whom died from the wounds suffered.[35] Finegan's Brigade, on June 3, 1864, distinguished itself in the so-called "Second Battle of Cold Harbor" when, just as General Breckenridge's lines were broken, they charged and recaptured the position and inflicted heavy casualties on the federal troops. Luckily for Hope's company, it suffered no casualties in this heavy battle, although the brigade lost fifty men to Union fire.[36] The brigade was stationed along the far right of the defenses and established entrenchments along the ridges near the Chickahominy River. Constantly under cannon and sniper fire the troops suffered greatly from the enforced inactivity and hot, dry Virginia summer. According to historian Gary Loderhose's history of the 9th Florida, the morale of the troops throughout June and July of 1864 was very low and desertion was openly talked about in camp. Many of the men believed that they should be back in Florida defending homes and family. Moved to Petersburg by June 19, 1864, the troops from the brigade suffered greatly from disease and boredom. For men of action, trench warfare was tough duty.[37]

According to an unnamed source in the files of the Pinellas County Historical Museum, Heritage Park, which was taken from the *Soldiers of Florida*, Hope was wounded on August 25, 1864, at Petersburg, Virginia. The wound must not have

---

[34]Loderhose, 60-64.

[35]Loderhose, 72-73.

[36]*Ibid*, 77-78.

[37]*Ibid*, 89-105. Loderhose has aptly entitled his sixth chapter, "Glory Fading." A recent account of the actions involving the brigade can be found in Zack C. Waters, "Tell Them I Died Like a Confederate Soldier," *Florida Historical Quarterly*, 69 (October 1990), 156-77.

been serious, since there appears to be no break in his active duty until October 1864, when he was elected to the legislature. This brief respite from the fighting did not last longer than a month and he returned to Virginia. He remained there until the last day of the war, surrendering his arms, like the rest of the Florida troops with General Robert E. Lee at Appomattox Court House. His discharge from the Confederate service was dated April 9, 1865, under Special Order Number 260.[38]

Sam Hope's election to the legislature, during the middle of the fighting around Petersburg and which included such noted skirmishes as Weldon Railroad, Cold Harbor, Reams Station and Hatcher Run, was not an uncommon occurrence during this war. Many of the legislators with whom Sam Hope served were also on leave from active duty. Sam's assignments in this extra-ordinary session of the wartime legislature included the chairmanship of the Internal Improvements Committee and membership on the Committee on Corporations.[39] During his stay, he introduced only one bill for the relief of Benjamin Hagler. He did not appear as a speaker during the debates or as an active introducer of motions, acts or resolutions. However his voting record was very consistent in constantly opposing any granting of discretionary powers to the Governor or county commissioners. He also voted against an attempt to require local troops to serve anywhere other than their immediate neighborhoods. He opposed an attempt to limit what might or might not be grown by individuals on their own land.[40] Late in the session, he was added to the Committee on Elections, after serving on a special committee formed to investigate the accounts of former governor Madison S. Perry and Quartermaster General of Florida H. V. Snell, his former commander and friend.[41] His attendance was excellent in not missing a single day of the term and he missed very few floor votes. Like every other member, he did not vote against anything

---

[38]Department of Military Affairs: Special Archives Publication Number 93. "Florida Soldiers: CSA 9th, 10th, 11th Florida Infantry," 211. State Arsenal, St. Francis Barracks, St. Augustine, Florida. This source lists Sam Hope's service as: Muster In (June 21, 62); Mustered Out (April 9 '65); Remarks (Wounded at Petersburg August 23, '64: resigned November 1, '64). The resignation was caused by his election to the legislature. Upon return to duty, he was restored to his previous rank.

[39]*Journal of the House of Representatives of the General Assembly of the State of Florida, 13th Session, 1864.* (Tallahassee: Dykes and Sparhawk, 1864), 33.

[40]*Ibid*, 96-107. During the debates covered by these pages, Hope showed his consistent voting against discretionary powers of the executives, state and local. This pattern, which was very clear, indicated his belief in the things that most southerners were fighting for when opposing the enforcement of national laws on certain issues believed to belong solely to the State.

[41]*Ibid*, 107, 55-56. The special committee did not find any wrongdoing on the part of Perry or Snell.

that might adversely affect the soldiers in the field or their families at home, such as limitations on what could be grown, pensions and widow's benefits. Upon completion of his term, he immediately reported back to service.

His return home did not mean the beginning of inactivity. In 1865, he was elected to represent Hernando County in the Constitutional Convention. At this futile convention, he served on the Committee on Militia and the Committee on Public Domain and Property and Internal Improvements, both natural assignments for a frontiersman who had served in the militia and as a Deputy Surveyor.[42] His voting record was not exceptional and generally reflected the feelings of the majority at this ill-starred meeting. The constitution passed by this assembly of men, deprived Blacks of the right to vote, petitioned the federal government for the removal of Black troops stationed in Florida and deprived anyone employed by the federal government, such as soldiers, sailors or tax-collectors unless they were qualified voters and residents of Florida, from voting or running for office. It also gave to the governor powers similar to those given to the president and presaged a strong centralized state government. It was exactly this type of document that drove many Radical Republicans to the brink and brought about the strongest measures of Reconstruction. Sam Hope did not sign the final document.[43]

Immediately after his return from the front, Sam Hope also participated, in a small way, in the escape of Judah P. Benjamin. According to Hope's account published in the *Confederate Veteran* in June 1910, Benjamin came to Hernando County and stopped at the residence of Leroy G. Lesley. Hope stated that he talked with the fleeing former cabinet officer while he hid out at Lesley's home. He did not, however, disclose the topic of discussion. From there, Hope relates, "Captain Lesley took him in his buggy to Braidentown, Mannatee [sic] County, to an old friend, Capt. Fred Treska, an experienced seaman. Captain Treska took charge of Mr. Benjamin and landed him safely in Bahama with a small sailboat."[44]

The years after the war brought Hope some additional family responsibilities, namely the birth of six additional children. The first two daughters, Susan Mary

---

[42]*Journal of the Proceedings of the Convention of Florida* (Tallahassee: Dyke and Sparhawk, 1865), 25-26.

[43]*Ibid*, 117. No document has surfaced to explain why Sam Hope did not sign this constitution, however, it would be within the realm of reason, given his strong dislike of centralized government, that it was the provisions giving the executive branch so much power that may have persuaded him to withhold his signature.

[44]*Confederate Veteran*, 8 (June 1910), 263.

and Grace May, were born prior to the end of the war, but on September 30, 1865, the first son, Samuel E. Hope, Jr., was born. After this blessed event, two more daughters and three sons were born, making a total of eight children. Sadly, when Sam Hope passed away, in June 1919, just three months after the death of his daughter Grace May, he had outlived all but two of his children, Clara Hope Baggett and John James Hope. His wife Mary lived until August 14, 1926.[45]

Sam Hope returned to an economy that was devastated by the war and offered few new avenues to wealth and security. However, he was always resourceful and soon entered the land business, both as a broker and surveyor of private properties. His relatively frequent letters to Hugh A. Corley, the mainstay of the land office in Tallahassee for nearly four decades, show a number of entries for lands in his and his family's name. The object in some cases was to secure homesteads for these members, however, because of their locations, some of these entries were probably for speculation.[46] The most notable cases of the latter type were those sections entered in the swamps of northwestern Hernando County, near the Chassahowitzka River. As these lands are too swampy for any useful homesteading, the speculation theme can be the only answer for their entry.

Hope, like many other men of means during this era of Florida's history, was speculating in lands rich with white cedar, the type used by the Eberhardt-Faber, Eagle Pencil and Dixon Crucible firms for the making of writing pencils. The fabled boom in this industry, centered at Cedar Key, Florida, is well known to most Floridians and need not be repeated here, except to note that the rapid growth of the industry and the heavy harvesting of these trees led to many charges of harvesting on state-owned lands. Hope was not immune from such charges. On June 20, 1877, Sheriff D. L. Hedrik, of Hernando County, wrote to Corley:

> I [have] written you Some time past in relation to H T Lykes and S E Hope. How mutch Land they had entered your reply was that Lykes had only Entered fourty acres: I wish to ascertain exactly how much H

---

[45] File, "Samuel Edward Hope," Pinellas County Historical Museum, Heritage Park, Largo, Florida. The file lists all of the birth dates, marriage dates and death dates of Sam Hope's family.

[46] Florida Department of State, Division of Archives and Records Management. Series 914, Carton 14. Hereafter, Florida State Archives, Record Group and Carton Number. Box 14 contains a number of Hope's letters to Corley from January 30, 1866 to December 22, 1872. The lands noted in these letters acreage due east of Brooksville, a section just west of Pasco (in Pasco County) and some very wet acreage in western Hernando County in today's Chassahowitzka National Wildlife Refuge.

T Lykes has Entered and allso S E Hope and William Hope. My object for making this enquirey is that they are cutting Cedar and I wish to do my duty in behalf of the State the maps I am in possession of says all is State lands Where they have been cutting pleas answer deffinately and as soon as convenant as the Cedar has not bin Shiped yet.[47]

Hope and neighbor and friend Dr. Howell T. Lykes invested in lands in this area specifically for the purpose of harvesting the cedar available there. Lykes' case became so bitter that he refused to negotiate a settlement with the local timber agent at Crystal River, C. T. Jenkins and, with many others who operated out of that cedar port, brought charges against Jenkins himself. Whether Hope was involved in this latter incident, is not known, although it would be difficult to see how he was not as the lands he did own were in juxtaposition with those Lykes was accused of abusing.[48]

For many years, Sam Hope was looking for a new place to call his own. His family was growing, his business interests took him farther away from the Brooksville area and communication was difficult. On August 12, 1877, he wrote the following to his friend, Hugh Corley:

Dear Hugh - I don't often think you make mistakes, but I think you did in regard to last letter you wrote me, you say the N.W. 1/4 of N.W. 1/4 of Sec. 28 Tp 22 R 19 was entered by Thos. H. Parsons there was such a man in this country long time ago but he is dead, and I have examined the Tax Books and his Exacutors does not give it in and it never has been claimed for him. Examine closely for me and be sure of it. Is the N.E. 1/4 of S.E. 1/4 of sec 34, Tp. 26. R 15. and N.W. 1/4 of S.W. 1/4 of sec 35 Tp 26 R 15 subject to Entry or not. By letting me know, you much oblige.[49]

This land was at the mouth of the Anclote River, on the north shore, and this letter indicated that Hope was very much interested in registering it in his name. This land was later included in the S. E. Hope Subdivision at Anclote, Florida.

---

[47] Florida State Archives, Series 914, Carton 19, Letter of June 20, 1877. Hedrik to Corley.

[48] See letters of December 16 and 29, 1879, in Florida State Archives Series 914, Carton 21 and Letters of February 3 and March 4, 1880, in Florida State Archives Series 914, Carton 22. Letters found in the correspondence of Timber Agent E. T. Berry, also from these cartons, also tell part of this story. The letters cited above are all from C. T. Jenkins to Hugh A. Corley.

[49] Florida State Archives, Series 914, Carton 19. Letter of August 12, 1877. Hope to Corley.

The next year, 1878, Sam Hope moved his family to their home by the river, where he was to remain until 1906.[50]

Throughout the remainder of his life, Sam Hope worked in the real estate business and occasionally did some private surveying. His pursuit of a comfortable life paid off very well and his home on the Anclote River attracted a number of people by the turn of the century. For many years prior to the founding of Tarpon Springs, the mail was delivered to the home of Hope's son-in-law, Joseph B. Mickler. The house still stands on the shores of the river and still has the slot through which the out-going mail was deposited when the Micklers were not at home.

Before and immediately after his move to Anclote, Hope served two terms in the State Legislature. The first term in 1874 found him on the Committee on Fisheries, chaired by the notorious William Gleason. His other assignment was on the Committee on Legislative Expenses.[51] This could not have been a pleasant session for Hope because it was one of the Reconstruction Legislature's and was filled with those who sympathized with the Radical program. Hope was always the self-declared conservative Democrat and never hid this fact from anyone. He hated the carpetbag supremacy that he felt controlled state politics, yet in 1874, he attempted to get some legislation through that would lessen its impact. On January 13, 1874, he introduced "an act to prevent Attorneys-at-law from acting as Clerks of Sheriffs or Deputies of either."[52] This obviously aroused a great deal of opposition and he was probably told that it had no chance of passage. The reason he took on the legal profession was that many of these gentlemen, almost all northerners, or like his old enemy James T. Magbee, turncoat scalawags, were acting as assistant sheriffs to newly enfranchised Blacks, many of whom never had the opportunity to learn to read and were being led, at least in the eyes of those like Hope, down the wrong path by these outsiders. He later had the bill withdrawn from consideration.[53] His only other attempt at legislation was the passage of a resolution to establish a mail route in Hernando County, which was passed on a voice vote.[54] In line with his conservative philosophy of government, he voted

---

[50]From copies of Subdivision plats in possession of Mr. and Mrs. L. E. Vinson of Tarpon Springs, Florida. Used with permission of the Vinsons.

[51]*Journal of the Proceedings of the Assembly of the State of Florida, Seventh Session, 1874* (Tallahassee: Hamilton Jay, 1874), 46.

[52]*Ibid*, 53.

[53]*Ibid*, 71.

[54]*Ibid*, 69.

against a bill requiring parents and guardians to educate their children and against a bill entitled "an act to prevent and punish Trespass upon the Public Lands of this State." In both cases he was in the minority and the bills became law.[55] He did not run again the following term, but, did submit his name for the session for 1879 and was elected.

In a campaign speech, following the Brooksville convention of the Democratic Party of Hernando County, he stated the following:

> It is not I as an individual that claims your votes, but as an exponent of Democratic and liberal principals, and the representative of that class of Citizens who further opposes tyranny, oppression, high tariff and Carpet bag supremacy. I have no political reputation which is tarnished, no accusations of turn coat to clear up, but as you all know my political principals have been purely conservative, and I now stand upon the broad platform, which was accepted at Cincinattis convention and Endorced at Baltimore. To the Colored voters I have to say in addition to what I have already said, I am your friend, Have I not shown by my actions, I challenge any one to say otherwise. If I am elected I cannot legislate for any laws for myself and not for you. The same laws that govern me will govern you and he that says to the contrary is both an enemy to you as well as myself.[56]

Hope is shown here as the conservative Democrat he had always been. The Cincinnati platform he alludes to, stressed a return to democratic principals and strongly urged, "Opposition to centralization and to that dangerous spirit of encroachment which tends to consolidate the powers of all the departments in one, and thus to create, whatever the form of government, a real despotism." It also emphasized "Home rule" and a tariff "for revenue only," things obviously dear to the heart of Sam Hope.[57]

The 1879 session was much more congenial to Sam Hope than that of 1874 and he landed a key assignment on the Committee on Railroads and Canals. This committee was chaired by John Westcott, another former surveyor and a major in the Florida 10th Infantry during the war. On the question of the expansion of internal improvements, such as railroads and canals, these gentlemen saw eye-to-

---

[55]*Ibid*, 320-21.

[56]Handwritten copy of Speech. From the files of Mr. and Mrs. L. E. Vinson, Tarpon Springs, Florida. Used with permission.

[57]Frank R. Kent, *The Democratic Party: A History* (New York: The Century Co., 1928), 269-71.

eye. Hope's first attempt at legislation was to get a resolution passed asking for a lighthouse at Anclote Key, which was passed unanimously by the House on January 16, 1879, early in the session.[58] Four days later, he pushed for a joint resolution to establish a mail route from Anclote to Tampa, via "Stevison's bridge."[59] Feeling that these resolutions would bring results, Hope next asked for another mail route, this one from Troy, Florida, to Anclote. This was passed with only one vote of opposition.[60] The remainder of this remarkable session, from Hope's point of view, was spent getting approval of numerous proposals for canals and railroads passed on for the governor's signature. The only other action requested by Hope during this term was Resolution No. 42 which was "an act authorizing the Governor to appoint a commissioner to adjust certain Indian war claims against the U.S. Government." No action was taken on the issue in the form proposed by Sam Hope, but in a more refined and improved act to examine and resolve these claims, passed as Assembly Bill No. 251.[61] These claims were one of the more important concerns of Hope's later life and something he felt deeply about.

Sam Hope had one more political function to perform before his active office seeking days were over and that was the election to the Constitutional Convention of 1885, where he again sat with Westcott, then the oldest member of this august body.[62] Representing Hillsborough County, Hope sat on the Committee for the Legislative Department and on the Committee on Enrollment and Engrossment.[63] A reading of the entire *Journal of the Proceedings* shows that Sam Hope was not one to introduce, at least from the floor, amendments or amendments to amendments. He was noted only once in the *Journal*, aside from voting, and that on a motion to kill any new amendments that had not first gone through the committee process. It was "laid over under the rule."[64] His voting record on this important

---

[58]*Journal of the Proceedings of the Assembly of the State of Florida, Tenth Session, 1879* (Tallahassee: C. E. Dyke, Sr., 1879), 71.

[59]*Ibid*, 79.

[60]*Ibid*, 107.

[61]*Ibid*, 227, which is his first proposal, and 423, which is the refined version.

[62]For more on John Westcott, see Joe Knetsch, "A Finder of Many Paths: John Westcott and the Internal Development of Florida," in Lewis N. Wynne and James J. Horgan, Editors, *Florida Pathfinders* (St. Leo, Florida: St. Leo College Press, 1994), 81-104.

[63]*Journal of the Proceedings of the Constitutional Convention of the State of Florida, 1885* (Tallahassee: N. M. Bowen, 1885), 53-56. The author would like to express his appreciation here to the staff of the State Library of Florida, Tallahassee, Florida, for allowing weekend checkouts of the House and Convention *Journals*. Without this cooperation, this work would be far less complete.

[64]*Ibid*, 156.

document showed, again, his conservative values and resistance to such things as high salaries for the governor. Because of the lack of letters home from this period and the form of the *Journal*, it is impossible, at this time, to determine Hope's exact role in the convention. In the stereotypical Gary Cooper mold of frontiersman, he probably took his colleagues aside and quietly persuaded them in his own fashion. But this is mere speculation and is undocumented.

Sam Hope had one other passion in his life that took many years and much of his personal time in advocating, though he never lived to see the final result. He was consumed by a drive to get the last of the Seminole War veterans paid their pensions and the other obligations honored that were promised in 1858. This struggle lasted throughout the remainder of his life and took him, many times, to Tallahassee to personally lobby the Legislature for the money. The first hurdle he did overcome, however, was an acknowledgement from the federal government that money was owed the State of Florida for the service of volunteer units. By the end of the 1870s, these claims were adjusted and paid by the Secretary of the Treasury to the State of Florida. The real struggle came with the payment by the state to the veterans of the Indian Wars. First the state had to pay the agents who procured the funds from Washington. S. I. Wailes, a powerful lobbyist and land agent, and W. K. Beard, of Tallahassee, acted as the agents and, in the end, received $25,000 from the account for the Indian War Claims as compensation for their activities. An additional amount of $132,000 was deducted from the Indian Trust Fund, which was paid back from funds meant for the Indian War veterans or their heirs. Although the two funds did not relate, the total amount, according to a typescript signed by Sam Hope was deducted from the veteran's money.[65] Hope was frustrated by the constant deduction of funds from the monies owed to the deserving veterans or their heirs.

In the early part of the new century, he privately printed a pamphlet on the topic showing the amount of funds, without adding any interest, due to the veterans or their families. Exhibit I of this pamphlet showed the total amount allowed to the troops after an auditor's report to be $163,645.79. Under a legislative act in 1861, the Florida obligated itself to pay these claims, yet, by 1902, when Hope and a select few received payments from the Legislature, most of the money had not been disbursed to the those deserving it. As Hope declared, "The state has held this money about long enough to turn it over to the proper owners. Most of

---

[65]Typescript of the letter sent to the Editor of the *Florida Times Union*, Jacksonville, Florida, n.d. Typescript from Mr. and Mrs. L. E. Vinson, Tarpon Springs, Florida. Used with permission.

these old soldiers are dead, but they have children and grand children and should be paid from the Muster as paid by the U. S. Government."[66] He tried nearly every avenue open to a private citizen to get the old soldiers paid.

One of the more interesting collections is the correspondence between Hope and his old friend and colleague, John T. Lesley. During the time of the correspondence presently available, the two men discussed various strategies to use on Governors Broward and Gilchrist. Hope noted that Broward had listened closely to what he had said and acknowledged the recommendations for appointments on the claims commission wanted by the veterans. Hope then wrote, "Now if you and Perry G. Wall will write to the Gov. and ask the necessity of having good men appointed on the commission you may help it along."[67] Later in April 1909, the claims still not paid, he again wrote to Lesley informing him that, "I wrote to Gilchrist a long letter he opened the way. And I give it to him right & left."[68] But, alas, this letter was to no avail and the pensions were still not paid at the time of Sam Hope's death in 1919.

It was on the frontier that Sam Hope made his name as an Indian fighter and officer. He pioneered a new settlement on the Anclote River and made it a permanent home for his family and many friends. His determination never showed truer than in the pursuit of the Indian Wars pensions that lasted well beyond his lifetime. His conservative principals remained with him throughout his life and reflected the lessons of that life on the wild frontier. He always resisted placing too much power in the hands of governors or county officials and opposed the Radical Reconstructionists' views of Florida and the South. His principals dictated that every man should be treated fairly and that each should have an opportunity to make the land do what it could for the benefit of their family and home. In the traditions of frontier democracy, Sam Hope best exemplified what it these values meant on the Florida frontier. The traditions of self-reliance, strong family bonds and the willingness to fight for principals deemed fitting to all were the hallmarks of the life of Sam.

---

[66]Samuel E. Hope, "Those Indian War Claims: A Full Truthful History," Privately Published n.d., Exhibit I, no page number.

[67]Letter of June 18, 1907. Hope to John T. Lesley. In private collection of Lesley family at the present time. My thanks to Kyle VanLandingham, who has examined this collection, for his working copy of the manuscript letter and the loan of it for use in this article.

[68]Letter of April 12, 1909. Hope to John T. Lesley. Private collection of Lesley family.

CHAPTER 14

IMPOSSIBILITIES NOT REQUIRED: THE SURVEYING CAREER OF ALBERT W. GILCHRIST

For surveyors, simple instructions can be deceiving. A basic order to follow along the line of "mean high water" in the meandering of a beach appears straightforward. It is clear, perspicuous and concise. What instruction could be easier to follow? How difficult could it be to determine such an obvious line? In the mangrove and buttonwood jungles of coastal southwest Florida, this simple one line directive could be one of the most difficult assignments given to anyone.

When he signed his contract and posted his bond in June of 1897, U. S. Deputy Surveyor Albert W. Gilchrist had little idea of the difficulties he was about to encounter. This was not because of a lack of knowledge of the terrain he was to traverse and measure. He was very familiar with Sanibel and Captive Islands and the surrounding outcroppings. Gilchrist had already surveyed lands on Gasparilla and LaCosta islands and had a good idea of the labor such a task would involve. What he was not prepared for was the extensive criticism he encountered from the Surveyor General of Florida, his superiors in Washington and some of the very settlers he was attempting to assist.[1]

Gilchrist had a great deal of experience in the profession of surveying. After receiving training in the subject at the United States Military Academy at West Point, he returned to Florida and found employment with the Plant System of rail-

---

[1] Gilchrist's contracts, bonds and some valuable correspondence are to be found in Drawer "U. S. Deputy Surveyors A-H," File, "U. S. Deputy Surveyor, Albert W. Gilchrist," Land Records and Title Section, Division of State Lands, Florida Department of Environmental Protection, Tallahassee, Florida. Hereafter, "Contracts and Bonds file."

roads. From 1882 through much of 1885, he was employed by the system in surveying routes throughout much of western peninsula of Florida. In the latter year he left Plant to join the staff of the Florida Southern Railroad, which had an agreement with the Plant System not to build a road to Tampa, but instead go further south to Charlotte Harbor.[2] By the following year, young Albert Gilchrist was settling down in the new frontier town of Trabue, named for its founder and benefactor, Issac Trabue. Trabue, an attorney from Louisville, Kentucky, had struck an agreement with the railroad to have the terminus of the line at his new town, which he had had surveyed and platted. But trouble soon arose and the inhabitants voted to change the name of the town to Punta Gorda, with Albert Gilchrist at first opposed but later voting with the majority. Setting up shop in the town, young Gilchrist soon had a thriving private business in surveying and real estate.[3]

While his business was beginning to take off, his political aspirations suffered their first setback in the first elections in Punta Gorda. Gilchrist ran for the office of mayor in the shadowy election of October 1887. W. H. Simmons, the town's first mayor, defeated him. One of the main reasons for this defeat may have been his hesitation to vote on the name change.[4] The loss of this office did not dampen his enthusiasm for office holding and he was soon to be back in the thick of local politics.

On December 6, 1887, he wrote to Surveyor General William Bloxham, the former governor, for permission to survey an uncharted island in Section 19, of Township 44 South, Range 22 East. In a rather unusual arrangement with Dr. Issac A. Silcox, his employer, he advised Bloxham that, "I understand he will be paid in land warrants, which he can turn over to me as payment for my services." This is the area now known as Josselyn Island, of the coast of Big Pine Island.[5] In

---

[2]Vernon E. Peeples, "Charlotte Harbor Division of the Florida Southern Railroad," *Florida Historical Quarterly*, 58 (January 1980). Peeples fully explains the events leading to the construction of the railroad to Charlotte Harbor. To date, no one has been able to uncover the documents, if they exist, which show why, how and when the deal was struck between the two systems.

[3]Vernon E. Peeples, "Trabue, Alias Punta Gorda," *Florida Historical Quarterly*, 46 (October 1967), 145.

[4]*Ibid*, 145.

[5]*Miscellaneous Letters to Surveyor General*, Volume 13, 388. Letter of 6 December 1887, Gilchrist to Bloxham. Land Records and Title Section, Division of State Lands, Florida Department of Environmental Protection, Tallahassee, Florida. This collection of bound letters is very extensive and is not indexed. All of the letters in the volumes are originals, however, because of their fragile nature, microfilm copies are used and open to the public. Hereafter, *MLSG*, volume number and page number (if available).

March of the following year, Gilchrist surveyed lands on Gasparilla Island and made some keen observations that put the Surveyor General's Office on notice that something was amiss. On March 25th, he wrote, "The plats at Gainesville, U. S. Land Office, shows the northern part of Gasparilla, unsurveyed. Section 4, T. 43 R 20 E is the most northerly point represented on the Island. Will you please send me a plat of the Island north of Section 4 in T 42 R 20, and the adjacent Islands...There are numbers of surveys down here requiring careful work. I would ask if you [illegible] not appoint me Deputy U. S. Surveyor." By which he meant, obviously, that he did want the position. Curiously, the Surveyor General's Office replied that, "Our maps do not show any Survey of northern part of Gasparilla Island in 42/20. Survey stops off at line starting between Township 42 & 43 S."[6] The Surveyor General, however, could not allow Gilchrist to make the surveys needed to answer the wishes of the inhabitants of Gasparilla or any other islands along the coast because the Commissioner of the General Land Office in Washington had issued a stop order for all further surveys of these islands. Gilchrist implored Bloxham to write the GLO and ask for a reconsideration of this policy. As he noted, "I would note that our country is rapidly settling up, owing to the advent of the R. R. There are numbers of Islands occupied by citizens, who are anxious to secure their land titles and are willing to pay for the survey. The action of the Commissioner in ordering no more surveys of Islands works an unjust hardship on our locality and Charlotte Harbor Bay in particular."[7]

The surveys of the islands in Charlotte Harbor would be a major challenge to the former Gadsden County resident who was born just three years prior to the outbreak of the Civil War on January 15, 1858, in Greenwood, South Carolina, his mother's family home. The son of the prosperous farmer, William E. Kilcrease—note difference in name's spelling—young Albert did not have the opportunity of knowing this strong-willed man for very long. His father passed away in May 1860 and the wealth he had amassed soon vanished as the Confederate dollar became worthless. Like most Gadsden residents, Albert and his mother, Rhoda, saw the residents of the wealthiest county in Florida sink rapidly into debt, depression and destitution. Only through the local political connections did Albert find his way into West Point, where he began the studies that led to the profession of

---

[6]*MLSG*, Volume 14, 66. Letter of March 25, 1888. Gilchrist to Bloxham.
[7]*MLSG*, Volume 14, 88. Letter of 16 April 1888. Gilchrist to Bloxham.

surveying.[8] For Gilchrist, who lived through Reconstruction Florida and who made a habit of taking up large challenges, the surveys of the rugged coastal islands of southwest Florida were to be one of his toughest of his life.

The islands had been reserved for possible military use prior to the Civil War and not until 1885 were some of them released for sale and homesteading. Supposedly the larger islands of Pine, Gasparilla, LaCosta, Sanibel and Captiva were surveyed in the early 1870s by Horatio Jenkins, a carpetbag politician who had prospered as part of the political machine in Duval County. His abilities as a politician far outstripped any he possessed for surveying. Modern surveyors up and down the west coast of Florida have constantly reflected upon the poor quality of his work. This frequent criticism is an echo of the words written by A. W. Barber, Examiner of Surveys, when he wrote, "the original survey of 1875 was grossly inaccurate and largely fraudulent; the section lines exist only on paper, by protraction."[9] Faced with having to deal with fraudulent surveys and nonexistent lines, Gilchrist was sorely tested as a professional surveyor.

He did not always have the cooperation of the established settlers in the region when he attempted to find the lines or survey their property. Some, like C. W. Wells of LaCosta, stated, "The was som surveyors down Hear Surveying and claimed tha Could not be any lines as proof to the Island ever bin surveyd and also got up a pertition for a resurvey of the Island Since then I have found the Lines and witness trees Thear is No use of a resurvey of the land a gain it will just cause Confusion and do no good."[10] Fellow surveyor G. H. Milman who homesteaded on Sanibel, confirmed the fact that the lines did not exist and wanted the islands resurveyed for proper legal descriptions. He, too, noted the lack of quality of the

---

[8]For the best study of Gilchrist's life to date, see Ric A. Kabat, "Albert W. Gilchrist: Florida's Progressive Governor," (Unpublished Masters Thesis, Florida State University, Tallahassee, Florida, 1987). Kabat's interest is in Gilchrist's political career and his work is very strong on this aspect of his subject's life. There is little mention of his career in the surveying profession. Pages 7-28 discuss the family background and his early education.

[9]"Examinations of Surveys," Record Group 49, National Archives, Washington D.C. "Field Notes of the Examination of Surveys (Florida)...as Examined by A. W. Barber. December 31, 1899-January 19, 1900." Hereafter, "Barber Examination." My comments concerning modern criticism of Jenkins' surveys comes from my seminars on the history of surveys and surveying in Florida. Many of my students, all of whom are professional land surveyors, have commented negatively and given first hand illustrations of the poor work. Especially helpful in this regard are the numerous discussions I have had with Mr. Jeff Cooner and his colleagues at Johnson Engineering of Fort Myers, Florida, whom I wish to acknowledge for their time, efforts and documentation.

[10]*MLSG*, Volume 14, 186. Letter of July 10, 1888. C. W. Wells to Bloxham.

Jenkins surveys, "I wish to call your attention to the fact that the survey of Sanibel Island (recently thrown open to homestead entry) in Tp. 46 S., Ranges 21, 22 and 23 E., Florida, made by Horatio Jenkins, Jr. D.S. in 1875 is utterly fraudulent and to the effect such fraudulent survey is having upon parties now seeking to enter homesteads upon this island."[11] Partisan politics also complicated any discussion of these surveys and Republican C. W. Wells accused Democrat John Crawford of helping his political colleagues in Gainesville reject his claims on LaCosta. He went so far as to blame local political pressure for forcing him off this land and resettled in Leesburg, where he anxiously awaited the return of the Republicans to power.[12] Gilchrist, a staunch and relatively conservative Democrat, might have been the surveyor alluded to in these tirades.

Albert W. Gilchrist did not just survey the islands of Charlotte Harbor. In the early 1890s, his work carried him to nearly every corner of DeSoto County and parts of neighboring Lee County. These were prosperous times in DeSoto County because of the discovery of pebble phosphate in very large quantities in the bed of Peace River, the largest tributary to Charlotte Harbor. Arcadia and the surrounding settlements took on aspects of boomtowns. Land was in great demand and with this came the demand for more surveys. Gilchrist was busy enough to open a second office in the Bank Building in Arcadia in addition to his Punta Gorda office in the Southland Block. The letterhead on his office stationary, obviously used for letters to potential buyers, noted that the total commerce for Punta Gorda in 1887 was $50,000 but it had, by 1893, risen to $2,000,000.[13] These figures represented the influence of phosphate mining and the shipment of ore out of the Florida Southern Railroad terminal in Punta Gorda. It also represented the continued export of cattle from the rich range lands of DeSoto and southern Manatee counties. Gilchrist and his various associates were in the middle of all of this activity.

During this period, Albert W. Gilchrist also platted out numerous settlements

---

[11]*MLSG*, Volume 14, 296. Letter of December 26, 1888. Milman to the Commissioner of the General Land Office, Washington D.C.

[12]*MLSG*, Volume 16, 305-06. One letter undated and the other dated August 21, 1890. C. W. Wells to "Surveyor General of Lands."

[13]Florida Department of State, Division of Archives and Records Management. Record Group 593, Series 914. Correspondence of the Secretary of the Board of Trustees of the Internal Improvement Trust Fund. Carton 51, Folder "G", 1894. Letter of October 26, 1894. Gilchrist to Louis B. Wombwell. This letter has the map/promotional information on the back. Carton 48 (Same Series), Folder "G" 1891, contains letters with the letterhead referred to in the text. This tells the location of Gilchrist's offices.

and subdivisions, including the Gilchrist Subdivision in Arcadia. According to the calculations of Ric Kabat, whose work is the most usable biography of Gilchrist we have, Gilchrist in 1891 alone sold 144 parcels of land and purchased seventeen others. In the course of his business as a real estate broker, he was involved in 1,105 transactions in DeSoto County alone, as evidenced by the large amount of correspondence found in the State Archives and the Land Records and Title Section of the Division of State Lands.[14] In 1893, he became involved with an attempt to charter and form the South American and International Railroad. John W. Whidden of Arcadia and James G. Gibbs of South Carolina, who was probably his stepfather or first cousin, assisted him in this venture. The scheme was to connect Charlotte Harbor with Columbia, South Carolina, through many interconnecting railroads.[15] However, this plan did produce any results. The continued growth of the land business meant that he had more than enough to keep him occupied.

Gilchrist also found time to serve in the State Legislature during this period. He was elected to the House of Representatives in 1893 and 1895, but was defeated for re-election in 1897, mostly because of his lack of sympathy with the plight of the farmers. He was a staunch conservative Bourbon Democrat and opposed the currency reforms proposed by the Silver Democrats and Populists. As a real estate agent, he believed in a hard currency, namely gold, and was not infatuated with the inflationary schemes to freely coin silver. While in the House of Representatives, he stood for closing of saloons on election days, indexing real estate deeds, requiring complete abstracts for all property and taxation for school sub-districts. In his second term, he voted for the railroad commission. Prior to his service in the legislature, he had served on the county board of health and, more importantly, had been appointed first colonel and then brigadier general of the Florida Militia. He held the latter post until 1901.[16]

Beginning in 1893 and little known to Gilchrist, the settlers on Sanibel had begun the process of getting their island resurveyed. In early March of that year, they sent a petition to the Surveyor General of Florida John C. Slocum requesting that the work be done. When they had not received a reply, one of them, T. H. Holloway of St. James City, wrote to Slocum noting "it is pretty hard to be com-

---

[14]Kabat, 30-31.

[15]Florida State Archives, Record Group 593, Series 914, Carton 51, Folder "G" 1894. Legal sized sheet with the charter and questions at the bottom. No date is affixed to the document.

[16]Kabat, 32-44.

pelled by law to live on a piece of land & not know where it is. I have build & made other improvements on what is supposed to be my land but there is no man living can say it is mine."[17] Within the next few years, they were to get their wish.

The gentleman assigned to the task of re-surveying Sanibel and Captiva was Albert W. Gilchrist. It is doubtful that on the day he signed his contract and bond, Gilchrist would have ever dreamed of the problems he was to encounter. He left his office on June 20, 1897, and headed to Sanibel by the morning steamer. Because of the well-known problems with the Jenkins surveys, he requested that the Surveyor General send him specific instructions as to how to start the survey. As he stated the problem to the Surveyor General W. H. Milton:

> As to the location of the lines as regards the points and natural features as shown especially on the Harbor side, there distances, I am satisfied, though shown to be regularly chained, are irregular. I have checked enough of this man's work on LaCosta and Gasparilla Islands to know this. If these points are to be fixed as the appear on the map, regardless of where they are but to come, it would pay me to traverse that side before I attempted to connect them by Section lines. I have hear that this was measured by counting the strokes of the oar. On some parts of La Costa this was not even done. You can easily verify this by comparing the map with the coast chart 175, the chart is accurate, the map is not.[18]

Gilchrist soon received the instructions, which may not have given him as specific instructions as he deemed necessary. He continued to note for the benefit of the Surveyor General the false and inaccurate nature of the Jenkins work. These warnings should have told that gentleman that many of the marks supposed to be on the land, including meander corners, simply were nonexistent. The signs were there for those who wished to read them, however, it appears that the Surveyor General was not one of that number.

On June 28, he left for Pine Island to commence the work there in preparation for continuing the lines on Sanibel. He immediately noted that corners were missing and bearings and distances not given. The notes for the meander and triangulation of Section 13 of Township 46 South, Range 22, he complained, were omitted. How could he successfully begin the survey if the lines and bearings

---

[17] *MLSG*, Volume 17, No page number. Letter of 22 May 1893. T. H. Holloway to J. C. Slocum.
[18] *MLSG*, Volume 21, 213-14. Letter of 20 June 1897. Gilchrist to Milton.

were not available? On July 6 he wrote Milton that he had made two triangulations from Sanibel to Pine Islands, but, that the bearings given by the old survey for these lines were wrong. "It is impossible," he said, "for a line to be run from this Island to Pine Island with such Easting and Southing or Westing and Southing. My only recourse is to erect the M.C. for Secs 13 T 46 S R 22 E & 18 T 46 S R 23 E. by the points of land." The erection of a new meander corner in place of those allegedly set previously was not within the scope of his instructions. He continued, in this letter, to complain of the lack of notes for the setting of these corners and the running of various lines from them.[19]

As if the problems of a lack of information were not enough, the rainy season began in the first part of July 1897. Writing to Milton on July 10th, Gilchrist noted:

> I am getting along fairly well. A storm is now raging, commencing night July 8. mosquitoes are of course not bad during the storm. One of my men expressed the status fairly at Pine Island when he said if you would swing a bucket around your head, it would be full of mosquitoes. The rainy season is certainly upon us. An ordinary rain we do not mind, but there is now a regular storm raining almost continuously for nearly 3 days with a howling wind. The work is tedious owing to the inaccuracy of original Survey. Only about 1/3 or 1/4 of these lines were ever run....The chances are that he crossed over in an open space and set the corners from the Gulf, avoiding the mangrove swamps. No posts where there is mangrove growth can be set at <u>mean high water</u> because <u>mean high</u> water is often a mile from shore, nor can any meander line be run at mean high water along the mangroves. It would take 5 men a week to cut a line in red mangroves bordering the low beach. Such lines can only be run on the edge of the water bordering the mangroves. As soon as the storm is over I shall run over to the Gulf Beach, cutting through the Island, down the Range line.[20]

Although very truthful in his description of the difficulty of running a line through red mangroves, the surveyor general was not satisfied. The instructions specifically stated that the line would be run at mean high water and that was that, no matter the difficulty in determining or cutting said line.

Gilchrist also saw that by continuing the line projected from Sanibel to Cap-

---

[19] *MLSG*, Volume 21, 214-7. The page numbers in this volume were changed, by hand, over time and appear to have begun in the middle of this letter. Letter of 6 July 1897. Gilchrist to Milton.

[20] *MLSG*, Volume 21, 8. Letter of 10 July 1897. Gilchrist to Milton.

tiva, there would be a "jog" in the line. He noted too that the Surveyor General had already anticipated this problem. He recommended that the survey be stopped at Blind Pass and not continued to the north on Captiva, LaCosta and Gasparilla. Bluntly stating his position, "So much of the survey as affects said Islands I would recommend be thrown away, as that can not be continued on Captiva, LaCosta and Gasparilla."[21] Upon finding part of the range line, the surveyor continued on toward the meander of the beach on the Gulf side and, unexpectedly, found the lines to be east of the best topographical features by nearly 3/4 of a mile. Finally, he again urged that the survey be stopped at the northern end of Sanibel, and not be carried over to Captiva. The only positive sign arising from the survey was the discovery of an abundance of sea grapes, which he thought, "would make a fine wine."[22]

Toward the end of August, Gilchrist had reported that the survey of Sanibel was nearly complete and that he had delayed sending his returns because of the press of other business and a two-week illness. The excess of time beyond the deadline he explained by the fact that an error in the original survey had thrown his crew off several miles. The errors in the meandering, he maintained, also caused the excess of mileage to be thrown into the next township, which meant that the entire line had to be re-run to discover the error. This error, once discovered, led to a re-running of the boundary line, which meant that G. M. Ormsby, an early Sanibel settler, was not in any defined township and his property descriptions were invalid. The misidentification of Palmetto Key also added confusion and delayed the finish of the survey. For all of this work and calculation, the surveyor put in a bill for $566.83, which was more expected.[23]

Confusion relating to the instructions for filling out the required forms and in the proper manner also delayed the final submission of the work for review. Again the Surveyor General questioned the surveyor closely on the existence of the former survey lines depicted on the earlier returns. And, again, Gilchrist pointed out the essential problem of the Jenkins surveys:

> I would state that there is now no line of the original Survey in existence and from all appearances and from the statements of the settlers, there never was any such line. In reference to all the work done under Jenkins' Contract, it is safe to say that no man, on Sanibel, Captiva, La

---

[21] *MLSG*, Volume 21, 18. Letter of 21 July 1897. Gilchrist to Milton.
[22] *MLSG*, Volume 21, 32. Letter of 4 August 1987. Gilchrist to Milton.
[23] *MLSG*, Volume 21, 51, 89.

Costa or Gasparilla Islands has ever seen an old government line on any of these Islands. I own much land on Gasparilla Island. I have worked on it for land lines, I have never seen one, and I do not know, ever, had I ever heard of a man who has ever seen any such line on it. I have worked on La Costa Island, I have seen several of its settlers. I know of no such lines there.

He went on to state with great certainty that no lines had ever been found by any of the settlers of those islands. He also complained, though tactfully, that the instructions to run a line from one island and continue it on the next through triangulation was risking imperfect lines and endangering the property of the settlers.[24]

When R. L. Scarlett became Surveyor General for Florida in December 1897, Gilchrist immediately tried to acquaint him with the problems of the Jenkins surveys and to justify his actions in running the lines. He also tried to inform the new man about the conditions under which a surveyor worked in South Florida during the summer season:

Being far off from communication and knowing the survey had to be finished up, I completed the work at my own risk. Besides, in the Summer time, with rains, myriads of mosquitoes and sand flies, mud 10 to 12 inches deep, then was the time to get them [illegible] it, while I was hardened to it. Instead of diminishing the amount, if there is any way of estimating what a hell, the foregoing combination will make, I hope the estimate will be increased by the addition of the "connecting lines," heretofore mentioned for which no estimate is submitted.

Significantly, Scarlett recommended that the amount submitted for payment be reduced. This response began to set the tenor for the next group of letters to pass between the two men.[25]

With his large number of investments and other commitments, Albert Gilchrist became impatient with the Department of Interior's slowness in payment. On February 5, 1898, he wrote to Scarlett requesting information about when he might be receiving even a partial payment for the work performed.[26] By the time Scarlett got around to answering this question, numerous settlers on Sanibel, including the Lee County Surveyor, were beginning to question the validity of the

---

[24]*MLSG*, Volume 21, 128. Letter of 8 November 1897. Gilchrist to Milton. This is the last letter between these two men. Milton was replaced by R. L. Scarlett before the end of December.
[25]*MLSG*, Volume 21, 162 1/2. (6 page letter) Letter of 7 December 1897. Gilchrist to Scarlett.
[26]*MLSG*, Volume 22, 207. Letter of 5 February 1898. Gilchrist to Scarlett.

work. The population of the islands began to rise rapidly and knowledge of where exactly the lines were was becoming crucial. Again the Jenkins surveys came to the forefront of the discussions. In a letter of August 2, 1898, Captain Sam Ellis wrote that he had been informed by J. Jenkins, Jr. of Tallahassee who had been one of the crew on Horatio Jenkins' survey that "the surveys were made about 28 years ago on Sections 23 and 24 on Tarpon Bay by simply counting the strokes of an oar allowing 3 ft. to each stroke, and the map shows the land one mile too far west. Where the map is marked land there is water, and where marked water there is land."[27]

By early September 1898, Scarlett was searching for Gilchrist. The exact location of the surveyor was necessary for him to make necessary corrections in the field notes and vouchers. However the chances of the Surveyor General finding him near Charlotte Harbor were slim, since Albert Gilchrist had enlisted in the Army to fight in Cuba and was, at that moment, at Guantanamo, Cuba. Gilchrist had enlisted as a private, but was quickly promoted to lieutenant and then captain. Scarlett, who thought Gilchrist was jumping ship to avoid having to make the corrections began taking a very belligerent tone in his letters to the surveyor. Scarlett was to the point of demanding that the officer abandon the Army to finish out his contract. From the hills below Santiago, Gilchrist replied, "Under no circumstances could I leave in the face of the enemy, if we had one, or in the face of an epidemic, never. I have waited nearly a year on your Department [through] no fault of yours or your deputy, and I trust you will wait on me, especially as I am here in the government employ."[28] Scarlett did not appreciate the pointed references to the slowness of the government's reply. When Gilchrist wrote him on December 18 explaining the nature of surveying in the mangroves and the near impossibility of chaining in the morass while wading in chest high water to find the line of mean high water, the only response, written on the face of the surveyor's letter, was, "Merits no response."[29]

Scarlett was furious with his deputy and was not about to have his bureaucratic power flaunted. The situation had reached an impasse. Into the foray stepped an old political acquaintance of Gilchrist's, United States Senator Sam Pasco of Florida. Pasco's letter to Scarlett, dated February 17, played down the alleged "objectionable" tone of Gilchrist's reply, and reminded the Surveyor General that the

---

[27] *MLSG*, Volume 22, 25. Letter of 2 August 1898. Samuel Ellis to Scarlett.
[28] *MLSG*, Volume 22, 79. Letter of 2 October 1898. Gilchrist to Scarlett.
[29] *MLSG*, Volume 22, 218. Letter of 18 December 1898. Gilchrist to Scarlett.

deputy was in the service of his country and therefore, "it is proper to grant all possible indulgence to those who have taken up arms in defense of the country during the period of war."[30] Pasco's gentle pressure may have been the turning point in the relationship between Gilchrist and Scarlett, especially after the Senator noted that he had contacted the Commissioner of the General Land Office, who was Scarlett's boss.[31]

By May of 1899, the furor seems to have passed and Gilchrist was back on the job, attempting to find the nonexistent lines on Captiva Island. To do this, he had to retrace the lines and "locate [them] where [they] should have been put." He requested the field notes for Captiva and began to get the corrections made.[32] Gilchrist made the necessary corrections by the end of June, 1899 and filed for the payment due him for the original work. Yet before final payment could be made, the government sent down an inspector, A. W. Barber, to investigate the situation. Barber brought with him a new camera to take pictures of the mangroves that had so much delayed the work and had been the point of contention in running the mean high water line. In his report, Barber noted the controversy:

> In his original returns he stated that he had to either run outside in the water or back in dense brush almost impassable, so he went outside. The Surveyor General was disposed to think this was inexcusable, and a sharp correspondence ensued over it. I assured the Deputy that the Department did not require impossibilities, and that the meandered shore line was not regarded as a strict boundary of the lots; and that his method would receive no further criticism...In conclusion, I cannot see how Dep. Gilchrist can have made any profit on this work, and he certainly has not intentionally slighted it in any respect. The accounting officers will only do justice to his good faith if the allow compensation for his patient searches and retracements, on a liberal basis. I think the survey should be accepted as now made; and that the Surveyor General should first be directed to make the lottings of the sections conform to the diagram furnished by the joint action of the Deputy and myself.[33]

Barber's report had worked wonders for Gilchrist's esteem and smoothed the way between the two men. Indeed, Scarlett even offered a contract for surveying

---

[30]*MLSG*, Volume 22, 53. Letter of 17 February 1899. Sam Pasco to R. L. Scarlett.

[31]*MLSG*, Volume 23, 41. Letter of 6 February 1899. Pasco to Scarlett.

[32]*MLSG*, Volume 23, 188. Letter of 28 May 1899. Gilchrist to Scarlett.

[33]Barber's Examination, 1. The photographs tell the story and show a nearly pristine Sanibel shoreline in 1900.

St. Andrews Bay to his former antagonist, but, because of other commitments, Gilchrist had to refuse.

After beginning his work in 1897, Albert W. Gilchrist, whose future as governor of the State of Florida still awaited him, finally got paid for his surveys of the islands in Charlotte Harbor. Writing to Scarlett on March 21, 1901, Gilchrist happily pronounced "I have re'cd the joyful news that I will be paid for the excess of my contract with the Gov. I feel very grateful. I thank you and the others who recommended it. Very Truly, Albert W. Gilchrist."[34]

It would be interesting to speculate just what the last thoughts of Governor Gilchrist were on May 16, 1926, when he breathed his final gasp, however, one may venture the idea that his surveys of Charlotte Harbor may have been among them. After all, he had attempted the impossible.

---

[34] *MLSG*, Volume 25, no page number. Letter of 21 March 1901. Gilchrist to Scarlett.

CHAPTER 15

HAMILTON DISSTON AND THE DEVELOPMENT OF FLORIDA

One of the most enduring mysteries in the history of Florida is the "man," Hamilton Disston. Little has ever been published about this individual's life or his many accomplishments. What is "known" is based upon few primary sources, and those have not been evaluated for accuracy. None of the few works that discuss the Gilded Age politics have ever delved into the life behind the man who bailed Florida out of its worst financial embarrassment. Even the story of his death is questionable, if, in fact, not totally erroneous. It is the goal of this paper to shed some light onto this unknown individual and maybe encourage greater primary research into background of those who have helped to shape our destinies.

That Hamilton Disston was a congenial person has been testified to by many who knew the young man. Born on August 23, 1844, he was educated at home, like many children of his day, and at the age of fifteen, he became a full-time apprentice in his father's factory on Laurel Street, in Philadelphia. His father was an inventive, strong willed and talented man, whose mechanical abilities came naturally through Hamilton's grandfather, Thomas Disston, of Tewkesbury, Gloucestershire, England. The adaptable Henry Disston migrated to America with his family in 1833, only to lose his father three days after their arrival. Left to his devices, he apprenticed himself to a saw-maker and began a career that reads like a Horatio Alger tale. With a start of only $350, he began his own manufacturing firm in 1840, and after some early struggles with land-lords and lenders, established himself at Front and Laurel Street, two years after Hamilton's birth. The elder Disston's skills could not be denied, and he looked to become independent of imported British steel. In 1855, he constructed his own steel mill, producing some of the highest grade crucible steel to be found anywhere. This gave him the edge over much of his competition, foreign or domestic. So successful was the Disston works, that they were not affected by the severe Panic of 1857. Young Hamilton, observing first-hand many of his father's administrative touches and inventive capacity, probably made mental notes of those that were most successful.[1]

The Civil War brought many changes to the Disston firm, primarily the need to produce war materials for the Union Army. For the purpose of making metal plates, whose importation was disrupted by Confederate raiders, Henry Disston constructed his own rolling mill. He also erected an experimental saw mill to test his saws on various types of wood for the purpose of more efficient cutting and precision. The father also experimented with new saw types, improved the quality of the steel and any number of improvements in various war materials. In the interest of the firm and national defense, he encouraged employees to spot defects and to suggest improvements.[2] The firm, aside from its primary business, also produced scabbards, swords, guns, knapsack mountings and army curbits for the military effort. In addition to his production, Henry Disston offered each of his employees who joined the colors half as much in addition to what the government would pay, and guaranteed them their jobs upon returning from the war.[3]

According to Harry C. Silcox, in his recent work on the Disston works, Hamilton wanted to enlist in the Army immediately upon President Lincoln's call for volunteers. Twice, Hamilton attempted to enlist, only to find his place taken by someone who was paid an enlistment bonus by his father, who insisted he was needed in the business. To relieve this tension, Hamilton increased his interest in the Northern Liberties volunteer fire company. As the story goes, the fires became so frequent that Hamilton was often missing answering the fire bell. Hamilton's popularity among this group of young men was great enough that, in the end, his father relented and allowed him and 100 "Disston Volunteers" to serve their time. The father went so far as to equip the entire unit. When the war ended

---

[1] See the following for accounts of Henry Disston: Herman L. Collins and Wilfred Jordan, *Philadelphia: A Story of Progress* (New York: Lewis Historical Publishing Co., 1941), 417-18; Allen Johnson and Dumas Malone, Editors, *Dictionary of American Biography*, Volume III (Cushman-Fraser) (New York: Charles Scribner's Sons, n.d.), 318-19; Harry C. Silcox, *A Place to Live and Work: The Henry Disston Saw Works and the Tacony Community of Philadelphia* (University Park: Pennsylvania State University Press, 1994); and William D. Disston, Henry Disston and William Smith, "The Disston History," May 1920. Typescript in the Disston Papers of the Tacony Branch Library of the Free Library of Philadelphia. The author would like to express his sincere thanks for the assistance rendered by this library's very capable and courteous staff.

[2] Johnson and Malone, *Dictionary of American Biography*. 318.

[3] "Disston History," 17.

in 1865, Hamilton returned to the firm and was created a partner in the new business of Henry Disston & Son.[4]

Assisted by a protective tariff policy and a growing demand for saws of all kinds, the years immediately following the war were very prosperous for the firm. New lines of products were introduced, including a new line of files developed during the war. According to a handbook from the company, "During the War we were unable to obtain files which would give us satisfaction and were compelled to manufacture our own. We spent thousands of dollars in perfecting our arrangements for manufacturing files."[5] Hamilton lost his social outlet when the volunteer fire department was discontinued in 1870 and this led him into his first ventures into politics. The prosperity of the firm and the free time that this allowed the "partner" meant a change of roles, one congenial with the growing need to market the company's new products and the move of the Disston works to the new area of Tacony.

Hamilton, like his father, was a strong Republican and favored the protective tariff. His growing interest in politics led him into the embroglio of the Philadelphia wards. In the beginning, he was allied with many of the so-called "bosses" of the wards, including James McManes (the city gas works "czar"), William Leeds and David H. Lane. He helped one of his old Northern Liberties Hose Company colleagues, John A. Loughridge, into the post of prothonotary to the Court of Common Pleas.[6] One-time governor of Pennsylvania, Samuel W. Pennypacker, noted in his autobiography, that, in 1875, Hamilton Disston was the ward leader in the Twenty-ninth ward, and was assisted by William U. Moyer. He also makes it clear that anything that went on in the ward, had to have Disston's approval, which he claims he disliked.[7] Pennypacker discribes his groups' defeat at the hands of Disston and his allies when he attempted to reform the precinct:

> "We hired a hall, notified every Republican, held a meeting which was largely attended and selected a ticket. For a time it looked ask though we would succeed, but we failed at the last moment through the better discipline of our oppo-

---

[4]Silcox, *A Place to Live and Work*, 54-55.

[5]"Disston History," 18.

[6]Silcox, 55. Samuel W. Pennypacker, in his *The Autobiography of a Pennsylvanian* (Philadelphia: John C. Winston Co., 1918), 176, noted that McManes had made his fortune in street railways. Pennypacker depicts this "thrifty, capable and vigorous Irishman" as an absolute autocrat who "tolerated no difference in opinion in the ranks." He states that McManes was the head of the Republican organization in Philadelphia during the 1870s.

[7]Pennypacker, *Autobiography*, 174.

nents and the superior practical knowledge which comes with it. The evening of the primary turned out to be cold, and blasts of snow filled the air. The well-to-do citizens upon whom we relied sat at home by their fires in comfort. Their servants rode in carriages, hired by the more shrewd regulars, to the polls and voted against us.[8]

The future governor learned from his tactical error and soon was on the way to more personal successes. Meanwhile, having learned how to control the ward, Hamilton shifted his efforts, somewhat, to the family-created settlement of Tacony, where he served innocuously as a Fairmount Park Commissioner, while controlling the town through his Magistrate (and real estate agent) Tom South.[9]

Disston's interest in politics also made him friends on the national scale. With an ability to help finance investment schemes as well as political campaigns, Disston had the ears of some of Pennyslvania's most powerful and influential. Among those who readily listened to and cooperated with Disston were Wharton Barker (once head of the Finance Company of Pennsylvania), Thomas Scott, Jr., the heir to the Pennsylvania Railroad former president's fortune and later, a partner in some of Disston's Florida ventures, and, most importantly, Matt Quay, U. S. Senator and "Boss" of that body. Through Quay, a high tariff man in his own right, Disston sought to keep the price of imported steel and, later, sugar, high.[10] He shared with his father, a strong feeling of getting things done, in politics as well as business. And, again, like his father, who, in 1876, served as a Hayes elector from Philadelphia, Hamilton attended to the political interests by attending the 1888 Republican National Convention as an at-large delegate.[11]

The Disston family also had investments other than their own saw-works. Hamilton's father, Henry, saw a need to keep ahead of the rest of the industry, and, as noted before, constructed his own rolling mill, an experimental saw-mill and

---

[8]Pennypacker, *Autobiography*, 174.

[9]Silcox, *A Place to Live and Work*, 55.

[10]For Disston's relationship to Wharton Barker, see Pennypacker's *Autobiography*, 124; For Scott's relationship, see *Florida Dispatch*, March 13, 1888. 218; For Quay, see, Stanley P. Hirshson, *Farewell to the Bloody Shirt* (Chicago: Quadrangle Books, 1968), 227-28. Hirshson also note Disston's ambition to run for the Senate in 1891.

[11]For his father's appearance as an elector, see, Johnson and Malone, *Dictionary of American Biography*, 319; For Hamilton's letter of June 18, 1888. J. J. Dunne to W. D. Barnes. "Old Railroad Bonds" (Drawer), no file (old box destroyed). Land Records and Title Section, Division of State Lands, Florida Department of Environmental Protection, Tallahassee, Florida. Letter is loose in the drawer at this time. It is on the letter-head of the "Keystone Chemical Company."

other smaller ventures near the family works. However, he also invested in a sawmill operation in Atlantic City, New Jersey, which helped to supply the saw works with handles for many of their tools.[12] Hamilton was more adventurous with his money and was an officer in the Keystone Chemical Company, the Florida land ventures and a railroad trust/syndicate, capitalized at $20,000,000, in China. This last venture was in association with Wharton Barker, Samuel R. Shipley (President of the Provident Life and Trust Company), and others.[13] It should also be pointed out that the Disston family owned much of the land in and around Tacony and a substantial sum was earned over the years from this investment.

As a manager of the Disston works upon Henry Disston's death in 1878, in his fifty-ninth year, Hamilton has received mixed approval and doubt. The firm's history, done by two members of the family and one other, states, "The general supervision of the establishment now devolved on Hamilton Disston, who possessed the quick eye and sound judgment of his father. He, together with his brothers...having served full apprenticeships in the shops, and with Albert H. devoting himself to the general financial and office management, was competent from a mechanical as well as a business point of view to carry along the intentions of the founder, and the steadily increasing business was pushed to proportions perhaps unexpected by him." The story is told, in this loosely official history, that when the plant was visited by President Rutherford B. Hayes, Hamilton showed the president a rough piece of steel, which, he stated, he would convert into a thoroughly finished saw before the entourage left the building. Exactly forty-two minutes later, he presented the president with a new 26-inch handsaw, engraved with Hayes full name upon it. It had passed through twenty-four different processes in that short period of time.[14] Another source noted that Hamilton was, "a keen, progressive executive. Under his management the business expanded materially."[15] Harry Silcox, in his study of the Disston Saw Works, states that Hamilton spent too much time away from Tacony to be any major factor in the firm's success, and that his brothers and uncle, Samuel, were more responsible for the growth.[16] Edward C. Williamson, in his tract on Florida Politics in the Gilded Age, paints Hamilton Disston as a Nouveau riche Philadelphia society clubman

---

[12] Silcox, *A Place to Live and Work*, 54-55.
[13] Silcox, *A Place to Live and Work*, 55; Letter-head on letter of June 18, 1888, cited above. [Disston is listed as "Vice-President" on this letter-head.]; and Pennypacker, *Autobiography*, 125.
[14] "Disston History," 63, 72-73.
[15] Collins and Jordan, *Philadelphia*, 418.
[16] Silcox, *A Place to Live and Work*, 56.

who defied his father's wish to keep his eye on the family business and went his own way, something like the prodigal son.[17] Which ever the judgment, it cannot be denied that Hamilton Disston was a figure of importance and controversy.

It is generally agreed that Disston's first trip to Florida came in 1877, at the behest of Henry Sanford, who had served in various capacities in Republican circles, including a stint as Ambassador to Belgium. According to one writer, the first attraction was the "lunker black bass" found in Florida waters. However, given the lack of subsequent reporting of Mr. Disston's fondness for fishing, even while a resident of Florida, this speculation may be questioned. What is clear is that Mr. Disston became very interested in the agricultural possibilities of the State, assuming it could find a way to remove the water that often covered the entire State south of Orange County. By 1879, he came to the firm conviction that the drainage of the upper portions of the Everglades, as he saw them, could be reclaimed from the morass of South Florida.

What was needed was a means to acquire title to the swamp and overflowed lands of the area and then bring in the equipment to do the job. Disston soon convinced fellow Philadelphians, Albert B. Linderman, Whitfield Drake and William H. Wright, along with William C. Parsons, of Arizona, and Ingham Coryell of Florida to invest in a corporation for the drainage of the swamp lands of southern Florida. By the shrewd device of not paying any cash, but assuming the expenses of the actual drainage, the new corporation, known as the Atlantic and Gulf Coast Canal and Okeechobee Land Company, was able to work an arrangement with the Board of Trustees of the Internal Improvement Trust Fund, the State's agency charged with the responsibility to encourage a "liberal system of internal improvements". The avoidance of payment was used to circumvent the obligations of the Trustees under the injunction placed upon them by Francis Vose and others.[18] Vose, and his colleagues, had invested in Florida railroad bonds toward the end of the Civil War and expected to be paid at par, plus interest, for their redemption. After numerous legal wrangles, a New York court placed the injunction on the Trustees which forbid them to sell or bargain away any State lands until Vose and friends had been paid in full. This forced the Trustees to seek large, corporate buyers for its land and effectually stopped using land as an enticement for railroad

---

[17]Edward C. Williamson, *Florida Politics in the Gilded Age: 1877-1893* (Gainesville: University Presses of Florida, 1976), 73.

[18]T. Frederick Davis, "The Disston Land Purchase," *Florida Historical Quarterly*, 17 (January 1939), 203-06. This short piece is still the best and simplest way to understand the Disston Drainage Contract and the Disston Purchase. Davis is emphatic about the two separate arrangements.

investment. By late 1880, the sum owed to Vose had risen to just over $1,000,000. Because the Internal Improvement Trust Fund was (and is) separate from general revenues and restricted in its mission to funding internal improvements, this injunction placed the State in a bind and virtually ended any major railroad construction for nearly a decade.

The drainage contract, which was signed by all parties on March 10, 1881, called for the company to drain and reclaim all of the swamp and overflowed lands in the area south of Township 24 and east of Peace River.[19] When 200,000 acres of land had been reclaimed by the company, the State would begin deeding alternate sections of the reclaimed swamp and overflowed lands to it. Thereafter, the State would issue deeds as the work progressed. The State would benefit by getting half of the reclaimed land, which would now be worth a great deal more, and the company would benefit by gaining title to the other half of the land. Hopefully, the land would sell, and at a price that would allow the company to recover all costs, with some profit left over for the investors. The corporation soon issued 600,000 shares of stock at $10 per share, and began to assemble the dredges that were to accomplish the task.[20]

The drainage contract received some national attention that was the beginning of a new era in the awareness of Florida by the print media of the day. On February 18, 1881, the New York *Times* reported:

> An immense transaction, involving the reclamation of 12,000,000 acres of land, or one-third of one of the States of the Union, has been undertaken by a company of Philadelphia gentlemen with every prospect of success...The project of reclaiming this wonderfully rich country has been talked of for years, and it has long been considered feasible by many noted engineers...The leading man in the enterprise is Hamilton Disston, a young gentleman of great business energy and ample fortune, and present head of the great saw-manufacturing firm of Henry Disston & Sons...Under the agreement already made with the State, the company is required to begin surveys within 60 days, and within six months to put a force equal to 100 men on the work, and continue as expeditiously as possible until it is completed. It is proposed to drain the land

---

[19] The original language of the contract stated Township 23, but this was amended later. The Peace River is what was meant, but looking at today's map, one finds Peace Creek (the original language) is a small stream, running east to west near Bartow, and flowing into Peace River, which is the current name of the water body meant by the contract.

[20] Davis, "Disston Land Purchase," 205-06.

by a canal from Lake Okeechobee to the Caloosahatchee River, which empties into the Gulf of Mexico. Another canal may also be constructed to the east, tapping the St. Lucie River, which flows into the Atlantic. These canals will entirely drain the swamp, and from ten to twelve million acres of the richest land in the world will be reclaimed. The company will receive for the work one half of the land recovered, and it is expected that this will largely repay all expenditure of money that may be made in the work...The entire property of the company is below the frost line, and there would be no such damage done to orange plantations as those in Northern Florida have suffered this Winter...Each share will carry with it the right to an acre of land. The stock will be put on the Philadelphia Stock Exchange, and is expected will be sold readily. Several applications for stock have already been made by prominent gentlemen in this city.[21]

Within a short time, after the negotiation of the purchase of these types of lands by Disston, international attention was to be drawn to Florida and the land boom of the 1880s was to begin.

On May 30, 1881, Governor William D. Bloxham announced to his colleagues on the Board of Trustees of the Internal Improvement Trust Fund, that he had, "gone to Philadelphia and had there entered into articles of agreement with Hamilton Disston for the sale to said Disston of four million acres of land at twenty-five cents per acre, and placed said articles before the Board for the action of the Trustees." On the following day, the Trustees accepted the deal and approved the articles of agreement.[22] The New York *Times* reported the sale on June 17, 1881:

> Philadelphia, June 16—What is claimed to be the largest purchase of land ever made by a single person in the world occurred today, when Hamilton Disston, a prominent manufacturer of this city, closed a contract by which he secured 4,000,000 acres of land from the State of Florida. This huge transaction has been in negotiation for several months, and its success was owing to the shrewd tactics on the part of the agents of Mr. Disston. The land acquired, a tract nearly as large as the State of New Jersey, was a part of the public domain of the State of Florida under control of the Board of Internal Improvement of the State.

---

[21]New York *Times*, February 18, 1881.

[22]*Minutes of the Board of Trustees of the Internal Improvement Trust Fund of the State of Florida*, Volume 2 (Tallahassee: I. B. Hilson, 1904), 500-01. [Hereafter, Trustees Minutes, Volume #, Page #]

> Owing to the recent improved value of the land of Florida, this property has been anxiously looked after by capitalists of New York and Boston...there were renewed efforts on the part of the New York people, who were backed by a well-known German banking house of that city, and the syndicate from London, headed by ex-United States Minister to Belgium Sanford and the Boston capitalists to buy the land...The tract is situated north of Lake Okeechobee, and is nearly all below the frost line...It is Mr. Disston's intention to at once begin an emigration scheme which will result in a very large addition to the population of Florida. To this end, he has already established agencies in several places in this country, and will at once organize bureaus in England, Scotland, France, Germany, Holland and Italy.[23]

In fact, the emigration centers had already been ordered and people of influence contacted. One of the principle negotiators for Disston with the Trustees was the venerable John A. Henderson, one of the most influential men in the State at the time. Ingham Coryell, whose contacts include General Sanford and James Ingraham, was also, no doubt involved in the laying of the base for this transaction, although this does need further investigation.

In the words of Bloxham's biographer, "No single event in Florida's history has equalled this one in economic significance."[24] The simple terms of the agreement required a first payment of $200,000 and the balance at agreed upon intervals. The entire amount was due on January 1, 1882. At the receipt of the first payment, Disston received a deed for 250,000 acres of land. Upon receipt of the first payment, which was immediately applied to paying off the Vose debt, railroad companies began lining up to get Trustees approval for their schemes. No fewer than ten such firms, who had all anticipated this sale, were waiting for the release of the lands destined to develop the lands of Florida. Among the most anxious was an Englishman, a member of Parliament, Sir Edward James Reed, whose interest in Florida was already well established and who was prepared, even at that moment, to purchase the ailing Florida Central Railroad. Disston made his first payment even before the due date and, by September 1, 1881, had made payments of $500,000, in cash, with the exception of $15,000 in coupons. About this time, Disston was negotiating with Sir Edward J. Reed for the eventual purchase of half of the 4,000,000 acres. These negotiations were very discreet

---

[23] New York *Times*, June 17, 1881.
[24] Kenneth R. Johnson, "The Administration of Governor William Dunnington Bloxham, 1881-1885," (Unpublished Masters Thesis, Florida State University, Tallahassee, Florida, 1959), 58.

and little record of their occurrence has been left to posterity.[25] By December 17, 1881, Sir Edward, as he was known in England, had entered into an agreement with Disston to handle his half of the purchase and all that was left was to get the Trustees to agree to their arrangement. On the following day, Disston addressed a letter to the Trustees explaining the deal:

> Sir Edward Reed, of England, as you know represents large Rail Road interests in your State. I have succeeded in interesting him more directly in South Florida. He will take Two Million acres of my purchase and as I understand him, at once arrange a Land Company in England which will rapidly send many purchasers of land to Florida. He will proceed to Tallahassee without delay in order to complete the transfer to him of his part of my purchase, and will bring with him full details of the arrangement between him and myself. He will make the remaining payments due on my purchase.[26]

Reed, of course, was not alone in the deal, and had already informed Disston of his contact with Dr. Jacobus Wirtheimer and other Dutch investors. Both Reed and Wirtheimer were involved in various railroads in Florida, including the Florida Southern Railroad and the Pensacola and Atlantic Railroad.[27] On January 17, 1882, Reed sent the Trustees a copy of the arrangement between himself, Wirtheimer and Disston.[28] Reed was later to ask for extensions of time to make the payments, but did complete same by December 26, 1882.[29]

The sale of so much land could not but arouse the ire of many of Bloxham's political enemies and some of the other interests who had lost out in the bidding. Politicians, pandering to the big vote, small money, medium to small farmer, screamed at the give-away to large corporations. Some noted the fact that foreigners were getting all of the choice land and leaving honest homesteaders the scrap

---

[25] Davis, "The Disston Land Purchase," 207-08.

[26] Letter of December 18, 1881. Disston to Trustees. Brown Rectangular Filing Box, "Disston", Land Records and Title Section, Division of State Lands, Florida Department of Environmental Protection, Tallahassee, Florida. [Hereafter, "Brown Box, Disston"] According to the Florida *Daily Times*, of December 23, 1881, the negotiations took place in New York during a series of "frequent interviews." Reed, for purposes of biographical background, was a member of Parliament, a noted civil engineer, a naval architect and consulting engineer for both the Russian and Japanese navies (ironically) and the chairman of the Milford Dry Docks.

[27] *Trustees Minutes*, Volume 3, 199-200.

[28] Brown Box, "Disston." Letter of January 17, 1882. Reed to Bloxham.

[29] Davis, "The Disston Land Purchase," 208.

land. One of the most visible critics of the sale was the editor of the Fernandina *Mirror*, the scholarly George R. Fairbanks. He noted that northern states supported education, improvements and other functions through the sale of their public lands and they did not let them go for 25 cents per acre. The negotiations for the sale, he maintained, were held in secret with no public input whatsoever. Only the extreme poverty of a misrun government could cause such a calamity as this sale.[30] The Tampa *Sunland Tribune*, also spoke out against the "give-away" because its editor, J. B. Wall, feared that Disston would select all of the good land in Hillsborough County and ruin any chance of attracting a railroad to the port city.[31] Another group who protested the sale was made up of squatters and new homesteaders. The former because their practiced way of life would now be seriously curtailed, the latter because the land company may claim the land before the required five years for a valid claim could pass, thereby depriving them a chance a good land.[32] Many of these complaints were very legitimate and real. The threat posed by the sale, did make life for some of these groups very uneasy.

Disston was well aware of the problems the sale would have regarding settlers and squatters. He understood going into the deal that legitimate settlers on the land should be allowed to purchase their land at the State price, $1.25 per acre, and agreed to the stipulation that these settlers had up to one and a half years to lay their claims before the land office to get this price. However, what displeased him most was the fact that the language used by the Trustees was that all settlers now upon the land could get this price. All settlers, Florida Land and Improvement Company Secretary Richard Salinger noted, "include[d] Squatters pure and simple." However, Disston agreed to this language because, "it would be better to include this class as it might lead to complications which it were better to avoid."[33]

As to the cooperation of Disston with railroad interests, who may or would have claims within the lands his agents selected, Disston, himself, pointed out his total cooperation with the South Florida Railroad Company and the Jacksonville, Tampa and Key West Railroad.[34] Disston seldom had any major disagreement with the railroad interests of the State, except W. D. Chipley, the aggressive vice-

---

[30]Williamson, *Florida Politics in the Gilded Age*, 75.

[31]*Ibid*, 76.

[32]Johnson, "The Administration of William Dunnington Bloxham," 60.

[33]Brown Box, "Disston." Letter of October 3, 1881. Salinger to Judge E. F. Dunne.

[34]Brown Box. "Disston." Letter of December 18, 1881. Disston to Trustees, previously cited. This file also contains hand-written copies of the agreements with the two railroads.

president and general superintendent of the Pensacola and Atlantic Railroad, who claimed some of the swamp and overflowed land in southern Florida should be reserved for his, and other, railroads.[35] Disston's aid was vital, in the end, to the completion of the railroad to St. Petersburg constructed by Peter Demens. It was Disston, who could not reach a final agreement with Demens, who introduced the sometimes cantankerous Russian to Phillip Armour, the Drexels of Philadelphia and financier Ed Stokesbury, and thus provided the means to the salvation of the Orange Belt Railway.[36] Disston, himself, later became the builder and part-owner of the St. Cloud and Sugar Belt Railroad, and always had an appreciation for the power of the rail. It should be remembered that Disston had, at first, contemplated a railroad to cover the entire area of the Kissimmee and Caloosahatchee valleys, which was to be built under the name of "The Kissimmee Valley and Gulf Railway."[37]

One of the most severe criticisms of the sale was the problem created by the designation of swamp and overflowed lands as those to be sold. Much of the confusion, then and now, comes from the definition of swamp and overflowed given in the Swamp and Overflowed Lands Act of 1850. The crux of the question revolves around the language which designated lands as swamp and overflowed if they were covered with water, all or part of the time, and thus made unfit for cultivation. If the land could be diked or leveed off and drained so as to make them useful for agriculture, then they could be granted to the states as swamp and overflowed. This definition also applied to a surveyable section of land, in which fifty percent or more of such a section were of this description. In fact and practice, this meant that many of the swamp land deeds would, of necessity, contain large portions of dry uplands. This fact was not lost on the Disston interests. On October 18, 1881, shortly after the purchase was made, Ingham Coryell wrote to Governor Bloxham suggesting that the State and the land company resolve the problem of high and dry land, before any public out-cry could take place.

---

[35] Florida Department of State, Division of Library and Information Services, Bureau of Archives and Records Management. Record Group 593, Series 665, Carton 1. Letter of April 14, 1884, Chipley to Bloxham.

[36] Albert Parry, *Full Steam Ahead: The Story of Peter Demens, Founder of St. Petersburg, Florida* (St. Petersburg: Great Outdoors Publishing Co., 1987), 13-16. See also letter of support, dated December 29, 1886. Disston to Governor E. A. Perry. "Old Railroad Bonds," [Drawer] Loose, file box destroyed. Land Records and Title Section.

[37] Old Railroad Bonds, [Drawer], "Atlantic and Gulf Coast and Okeechobee Land Company" [File]. Letter of June 8, 1888. Disston to Governor E. A. Perry, Land Records and Title Section.

> I now propose on the part of the Land Company, "That we select by personal inspection all the high & dry lands within the limits of the reclamation district so soon as the condition of the waters will allow it, and that the State join us in the inspection by the appointment of an inspector to represent the State, and all lands returned as not subject to overflows be taken by the Land Company...to accomplish the object of setting at rest finally & forever any contention as between the State & our reclamation company as to what <u>are</u> & what <u>are not</u> reclaimed lands, would be pleased to have them do so. Some mode of settlement at this stage would perhaps avoid a vexed questioning in the future. We now have all the evidences of what <u>are</u> or <u>are not</u>, overflowed Lands & plenty of reliable Witnesses to prove the fact. If affected by drainage, the proof would not be so satisfactory.[38]

However, the Trustees did not respond to this suggestion. Problems, as predicted by Coryell, did, indeed, plague the project from this very lack of definition. It is no surprise that most land historians consider this act one of the most ill-designed land acts ever passed by Congress or administered by the General Land Office.

But, regardless of the criticisms, some did appreciate and defend the Disston Purchase. One of the most ardent of these defenders was Charles E. Dyke, editor of the Tallahassee *Floridian*. This newspaper, as it was known at the time, was the organ of the State's Democratic Party. Yet, it does not diminish the importance of his views, when it is known that most of the registered, white voters were Democrats. After laying out the main features of the purchase contract, Dyke wrote:

> As we have heretofore stated, the sale has our heartiest endorsement as a financial transaction in every way meeting the peculiar exigencies of the case. Indeed, under all the circumstances, it is difficult to see how the Board could have done otherwise to accomplish the purpose in view. The debt against the Fund already amounted to some $900,000, and was constantly increasing, in effect bearing compound interest. The expenses of litigation, added to the interest, annually increased the debt, notwithstanding the large sales of land and the application of the proceeds to its payment. The natural and inevitable result would have been the entire absorption of the lands belonging to the Fund in a few years, leaving nothing for internal improvements. Besides, all this, the

---

[38]Brown Box. "Disston," Letter of October 18, 1881. I. Coryell to Governor W. D. Bloxham.

creditors, cognizant of these facts, were preparing to apply to the courts to have the entire Fund and all its interests turned over to them, or to have some five or six millions of acres set apart and deeded to them in satisfaction of their claims. Had the Trustees remained quiet and allowed this to be done they would have been justly subject to censure.[39]

This defense of the Trustees' action was taken up by many of the other newspapers throughout the State and helped, in some measure, relieve some of the pressure on the Governor and the company.

As to the actual work of the drainage company, it began with two dredges, one beginning the work of connecting East Lake Tohopekaliga with Lake Tohopekaliga, and the other began work on the canal which would open Lake Okeechobee to the Caloosahatchee River in South Florida. The Trustees were notified of the commencement of work on November 7, 1881, by the president of the Atlantic Gulf Coast and Okeechobee Land Company, Samuel H. Grey, who had replaced William H. Stokely. Hamilton Disston, was the treasurer of that company, but president of the Florida Land and Improvement Company, which handled the land sales transactions for much of the Purchase lands. The success of the reclamation project was noticeable in a very short time. Francis A. Hendry, one of the early promoters of the drainage project, told the Florida *Daily Times*, in early 1882, that the work on the southern end of the project was begun in earnest on January 20, 1882, and had resulted in a canal 28 feet wide and 5 feet deep and one mile in length. This short canal had already shown evidence, according to the Captain, of an increased velocity of the water headed toward the Gulf and had begun to scour out a deeper and wider channel on its own. This, he believed, would eventually result in a large canal that would help to lower Lake Okeechobee some six feet and expose some one million fertile acres to agricultural use.[40] The northern work experienced one of the most famous incidents in the history of the drainage project.

> On November 22nd, [1883] the last dams on the line of canal were cut, and vent given to the waters of the lake. A number of visitors assembled to witness the interesting event. The first rush of the waters carried away the last vestige of the dams, and accumulations in the canal, and

---

[39]Tallahassee *Floridian*, June 28, 1881, 2.
[40]Florida *Daily Times*, February 23, 1882, 3.

the velocity of the current established was sufficient to scour out the softer strata composing the bed of the canal, to a depth several feet below the line of excavation...During the first thirty days, the lake surface fell thirty-six inches...East Lake Tohopekaliga, formerly surrounded by cypress and marsh margins, has developed a beautiful wide sand beach, the bordering lands are elevated and marshes changed to rich meadow lands.[41]

The Southport canal was the next to be completed in the northern area, connecting Lake Tohopekaliga to Lake Cypress. This canal cut off the tortuous channel, now called Dead River, and greatly shortened the length of time to ship goods southward from Kissimmee City, which was founded as a result of the company's efforts. The city of Southport, where the canal leaves Lake Tohopekaliga, was founded within a year of the canal being opened, so much had the level of the big lake fallen. Narcoosee and Runnymede were also founded on reclaimed land in the area and settled by English colonists.

Most importantly for the direct future of Florida was the establishment of the St. Cloud Sugar Plantation. Because Pat Dodson has written so well on this topic, I will only summarize its accomplishments here. After the reclamation of some of the land near Southport, an experimental patch of 20 acres of sugar cane were planted there in land that was covered with muck and two to three feet of water the year before. The results, monitored by Captain Clay Johnson and John W. Bryan for Mr. Disston, were astonishing. Rufus E. Rose, Clay Johnson's brother-in-law, and later State Chemist, began planting on a larger scale. Disston, in 1887, personally bought half interest in the St. Cloud Sugar Plantation and increased its capital so as to allow the planting of 1,800 acres. The result was a record harvest and yield, higher than any recorded in the United States to that time. Disston soon brought down contractors to erect a sugar-mill, costing nearly $350,000. The mill had a capacity of producing 372 tons per day, much above the average of 200 tons elsewhere. The sugar produced by the plant, in spite of the lack of sophisticated machinery, was excellent and profitable. Congress, in 1890, to aid in the domestic production of sugar, helped the enterprise along by paying a bounty of 2 cents per pound. As Dodson noted: "Influenced by the bounty, advice from sugar experts, and by increasing consumer demand for sugar, Disston took in more associates, and reorganized the plantation under the name of the Florida Sugar Manufacturing Company. It was capitalized at a $1,000,000, and an additional 36,000 acres were

---

[41] Elizabeth Cantrell, *When Kissimmee Was Young* (Kissimmee: Self Published, 1948), 25.

added to the holdings." Financially, this investment did not pan out well, even though the production was high. The panic of 1893 played a role in the loss in this investment, but, more importantly, the bounty so gratuitously given by Congress in 1890, was removed in 1894 by the Cleveland administration. With late 1894 and early 1895 came the freezes that so destroyed the citrus industry, and with it, land prices, upon which the whole operation depended, became ridiculously low. Although things picked up in 1895, it was not sustained long enough for Hamilton Disston to realize any major profit before his death in April of the following year.[42]

Other problems had an impact on the whole scheme of the drainage and land sales. For the drainage project, the problem existed of the amount of land actually reclaimed and deeded to the company. In spite of two very favorable inspections by state appointed engineers, the legislature, in 1885-86 ordered an investigation into the claims of the company and the commission appointed by the Governor found the company had exaggerated its reclamation efforts and that it was not entitled to all of the lands it claimed as a result of its efforts. The political nature of this commission can be seen in the appointment of J. J. Daniel, of Jacksonville, a highly skilled surveyor and attorney, who forthrightly wrote to Governor Perry, informing him that as a president of a railroad company, with interests, as an attorney, in other such firms, he was technically not qualified to be on the commission according to law. The Governor overlooked these problems and appointed him, along with J. Davidson of Escambia County and Col. John Bradford of Leon County, as commissioner.[43] Indicative of the tenor of the investigation, Daniel wrote to Perry:

> Dear Gov. After consultation with my associates, I write to say that we do not consider any of the lands lying South of the section line which runs two miles North of and parallel to the township line between townships 27 and 28 South, as reclaimed by the work of the drainage company. There are lands within the drainage district North of this line around Lake Gentry and Alligator Lake and in the water-shed of Reedy Creek which have not been effected by the lowering of the waters of Tohopekaliga La...we have not carefully examined the lands around Lake Rosalie and Walk-in-the-Water and it may be that there has been a

---

[42]This is a quick summarization of Pat Dodson's article, "Hamilton Disston's St. Cloud Sugar Plantation, 1887-1901," *Florida Historical Quarterly* 49 (April 1971), 357-369.

[43]Florida State Archives. Record Group 593, Series 665, Carton, Letter of December 4, 1885. Daniel to Perry.

partial reclamation effected here, though from the examination we made below, that is around Tiger Lake, we are not disposed to thing that the waters of Rosalie and Walk-in-the-Water can have been very materially reduced.[44]

Although later reports from Daniel indicated that the drainage had some impact on the removal of water from the land in the northern area, the company was forced to reconvey some lands already deeded to it and modify some of its operations in order to reach the magic 200,000 acres required by the drainage contract. Throughout the ordeal, Disston and his engineer, James Kreamer, maintained that the company had done exactly as it claimed.[45]

By March of 1889, Kreamer was reporting to Governor Fleming that the progress was becoming more rapid and that many of the promised canals had been dredged. His report of progress listed the following canals as being totally or partially complete by December of 1888: The Cross Prairie Canal, the Southport Canal, the canal connecting Lake Cypress to Lake Hatchineha, the connector between Hatchineha and Lake Kissimmee, the improvement of Tiger Creek and Rosalie Creek (called "Cow Path" in the report), improvements in the Kissimmee River itself, the canal between Lake Okeechobee and Hicpochee, the Hicpochee to Lake Flirt canal, and the canal from Lake Hicpochee southward into the Everglades. Also included in this report is the additional canal into the upper Caloosahatchee River and some improvements in this river's channel. Finally, there is the widening of the Southport canal to a width of 106 feet.[46] There is ample evidence that, in addition to lowering the levels of Lakes Tohopekaliga, East Lake Tohopekaliga, Lake Cypress and some of the smaller lakes to the east of this group, that the level of Lake Hatchineha was reduced. Enough of this lake was lowered to expose and area known as Live Oak Island, on the northeastern shore of the lake, approximately 182 acres in area, not counting the marsh.[47]

In 1893, as required by law, the Trustees issued another "Official Report...To the Legislature of Florida Relative to the Drainage Operations of the Atlantic and Gulf Coast and Okeechobee Land Company." This report noted that, up to 1893,

---

[44]*Ibid*, Letter of March 23, 1886. Daniel to Perry.

[45]*Ibid*, Folder 2. Letter of June 19, 1886. Disston to John A. Henderson.

[46]*Ibid*, Letter of March 5, 1889. Kreamer to Fleming.

[47]See a series of letters in Volumes 20-22 in the Miscellaneous Letters to Surveyor General. Land Records and Title Section. For the acreages involved here, see the Official Plat of Township 28 South, Range 29 East, surveyed by William H. Macy, January 16, 1897.

the company had been conveyed 1,174,943.06 acres of land. It stated that the canals mentioned in Kreamer's 1889 report had been deepened and snagged, thus facilitating more outflow of water from the land. This report also recognized the newer canals in the northern area to Lake Hart and from that waterbody to the Econlochatchee River. Finally, it claimed that the level of the great Lake Okeechobee had been lowered four and one half feet below its normal level at the time the contract was entered into. All in all, the report was very favorable to the company's interest and promised little trouble in the future deeding of lands under the old drainage contract.[48]

As for the settlement portion of Disston's work in Florida, one has only to look at the map around the drainage contract to see the towns of Narcoosee, Runnymede, St. Cloud, Southport and the city of Kissimmee to note the growth of this area. However, less publicized, though just as important, was the development of Tarpon Springs, Gulfport, Anclote, and other cities in modern Pinellas County to see an even greater effect on the growth of Florida. The main impetus to the growth of this vital area was through the Lake Butler Villa Company, another of the many Disston land companies. According to Gertrude K. Stoughton's *Tarpon Springs, Florida: The Early Years*, Disston and former Arizona governor, Anson P. K. Safford, were looking for a place from which to make a resort. James Hope, son of Anclote pioneer and U. S. Deputy Surveyor, Sam Hope, acted as their guide. Upon reaching Spring Bayou, the two immediately agreed that they had found their spot. So thinking, the Florida Land and Improvement Company was assigned title to about 70,000 acres of land in the vicinity, of which is soon transferred most of it to the Lake Butler Villa Company, of which Safford was the president. Disston also had agents looking further down the Pinellas Peninsula for additional opportunities. They settled upon the land that soon would be called "Disston City," today's Gulfport. Tarpon Springs flourished under the guidance of Safford and Mathew R. Marks, another experienced real estate man, who made a name for himself in Orange County before migrating west. Disston City, on the other hand, suffered greatly when Demens ended the Orange Belt Railway in St.

---

[48]"Official Report of the Board of Trustees of the Internal Improvement Trust Fund to the Legislature of Florida Relative to the Drainage Operations of the Atlantic and Gulf Coast Canal and Okeechobee Land Company, 1893" (Tallahassee: Tallahassean Book and Job Office, 1893). It is worth noting that the lowering of the Lake Okeechobee by four and one half feet, via this operation, would be greatly disputed today, as it was in the years immediately preceding Governor Broward's great effort.

Petersburg, named for the Russian's home town.[49] Thus, two very important areas in the State of Florida owe their very existence to the efforts of Hamilton Disston.

Also owing much to the Disston heritage is the Florida sugar industry. Although some sugar has almost always been grown in Florida since Spanish times, Disston's experiments in St. Cloud and the surrounding area proved the potential for the exploitation of the land for the growth of sugar. Also experimented with in this vicinity was rice. A separate company was founded to exploit this crop's potential also, but it was fairly short-lived compared to sugar.[50] Like other Floridians, he experimented in a variety of crops, including oranges and tobacco.[51] This latter would seem to be a natural product for the heavy cigar smoking Disston. His drainage idea inspired many to do the same with their lands and helped to bring about a whole new way of looking at swamp lands.

Little more needs to be said about his immigration schemes. He was successful in helping to bring a number of English settlers to Central Florida, where their heritage lives on today. He also brought in Italian laborers to work on the dredges in the early years of the drainage project. At one point, as a humanitarian gesture, he offered forty acres of land to each of fifty Jewish families displaced by the recent Russian pogroms who were stranded in Philadelphia.[52] The exact number of immigrants he brought to Florida is impossible to guess, but it was substantial.

Hamilton Disston was a complex man, like any of us. However, there has been a persistent story that needs further examination, and that is the alleged suicide of this active and vital man. The main sources of the story, and there are only two cited, are; one, the Democratic newspaper of Philadelphia, and; two, the oral testimony of a nephew who barely knew "uncle Ham," but stated that it was the family secret. Upon examination, one should remember the times in which this newspaper story was written; the age of "yellow journalism," William Randolph Hearst and Joseph Pulitzer. Remember, also, that Hamilton Disston was a large and reg-

---

[49] See Gertrude K. Stoughton, *Tarpon Springs, Florida: The Early Years* (Tarpon Springs: Tarpon Springs Area Historical Society, 1975, Second Edition) and Parry, *Full Steam Ahead*, 23-24.

[50] *The Tropical Sun*, (West Palm Beach/Juno) April 1, 1891 and June 10, 1891. It was reported in this latter article that 15,000 acres was rented for four years at a rent of $600,000. The firm was known as the Kissimmee Rice Manufacturing Company.

[51] Florida *Times-Union*, January 21, 1886. It was reported that Disston had 56,000 young orange trees planted in a grove on East Lake Tohopekaliga. For the tobacco story, see Florida *Times-Union*, December 4, 1895.

[52] *South Florida Journal*, March 9, 1882. He wrote the offer in an open letter to Mayor King.

ular contributor to the Republican Party, who also had, at times, some political ambitions of his own. Additionally, we should not forget that he was close to Matt Quay, the hated Republican "Boss" of the United States Senate, from Pennsylvania. The nephew's statement can be seen, in some circles, as hearsay. The alleged cause of the suicide was the supposed losses by the Disston firm during the Panic of 1893-94 and the coming due of a $1,000,000 note drawn on the family business.

The facts of the case do not lead to a conclusion of suicide, especially since there were no eye-witnesses and the coroner's official report, recognized by all, reads that he died by natural causes, probably a weakened heart, i.e. a heart-attack. The "Father of Business History", Alfred Chandler, in his book on Land Titles and Fraud, noted, on page 493, that he was acquainted with Hamilton Disston and that he died of a heart-attack, while at home. If the obituary in the May 1, 1896, edition of the Florida *Times-Union* is typical, the newspapers reported that, "Heart disease was the supposed cause of his death." How could only one newspaper get the story right, when a coroner, a friend and all of the other newspaper organizations in the country reported his death incorrectly?

Given the Victorian era's fondness for protocol, the funeral of Hamilton Disston has a place in this story. If someone of the stature of Mr. Disston had committed suicide, it is doubtful that any truly notable individuals would have attended the funeral. Employees of the deceased would hardly have been permitted by the company to attend such an affair either. In the case of Hamilton Disston's funeral the attendee list reads like the Philadelphia social directory and included Mayor Warwick, P. A. Widener, William Elkins, Thomas Dolan, State Senator Charles Porter, Governor Hastings of Pennsylvania, Mrs. Mathew Quay, District Attorney George Graham, etc. The Atlanta *Constitution's* listing, from which this is taken, noted that some were honorary pallbearers and those that were actual pallbearers. This newspaper's May 5, 1896 account also restates the cause of the death as heart disease. During this era one does not attract attention to the death-by-suicide by having such notables attend the services along with over one thousand of the workers from the family-owned factory.

What of the charge of near bankruptcy caused by the Panic of 1893-94? Yes, the Disston works did reduce wages by 10 per cent during the middle of the panic, however, the company was on sound enough ground that by May 23, 1895, they had restored the lost percentage and the firms business had increased 12 percent from April of 1894 to April of 1895.[53] On April 4, 1896, the Disston firm announced in the New York *Times*, "Value of annual product, $2,500,000. Our foreign trade is 20 per cent of our total business. Our output is 20 per cent greater that six years ago." This would hardly put a business out of commission. Finally, there is the will of Hamilton Disston. This document, as recorded in the New York *Times* of May 9, 1896, stated that: "In the petition filed by the executors the value of the estate is given as 'over $100,000,' but it is thought that it will amount to several million dollars when the heavy insurance Mr. Disston carried is included." The "income" from the will and its enjoyment was to go to all the children, until his son reached his thirtieth year, when he could then take his full one-third share. His wife was to get most of the material goods (house, horses, carriages, etc.) and enjoy all of the other income until her demise, when it would residuary estate. Fully one-third of the real estate, except that in Florida, was to go to the wife during her natural life time. This is not a document of a poor, destitute man, driven to the brink and beyond by a $1,000,000 note due on his investments. It is interesting to observe that his Florida holdings are excluded from his personal will. The answer why this is so can be no simpler that the fact that corporations, under law, are treated as separate, corporal bodies, with lives of their own. With Disston's friends in the financial and political worlds, there are few questions to be raised as to whether or not he could have won an extension, refinancing, etc. of this note. The entire scenario of a financially distraught, no place to turn man, bent on suicide, simply does not fit the available evidence at this time. The newspaper account of the suicide in a bath tub is much too melodramatic. The whole repetition of the story smacks of another "Seward's Folly" myth of American history.

In summation, Florida lost a good friend when Hamilton Disston passed from this world. This was recognized by his contemporaries. In its editorial for May 1, 1896, the day after Disston died, the Florida *Times-Union*, stated: Floridians will read the news of the sudden death of Hamilton Disston with a feeling of genuine regret. He did wonders for the advancement of Florida's interests and the development of her products. He can be classed as one of Florida's best friends." Kissimmee, St. Cloud, Narcoossee, Tarpon Springs, Gulfport, Runnymede, Fort Myers, LaBelle, Moore Haven, etc. all owe a debt of gratitude to Hamilton Disston. The great attempt to rescue Florida from the swamps, mosquitoes and alligators, and

---

[53]New York *Times*, May 23, 1895, 1.

make it a showcase of civilization is the legacy of Disston's Florida efforts. Without his leading the way, how long would the State have to await one like him: a Flagler, Chipley, Plant, all followed his lead and made their own marks upon the landscape. There may be those who chide these remarks and look at the environmental damage done by his and succeeding generations, however, they take the man out of his time, and thrust upon him a knowledge he did not have or have a chance to acquire. Without him, many who now criticize, would never have migrated to the Sunshine State and stayed away, wondering how anyone but the brave Seminoles could live in the land of swamps and alligators.

CONCLUSION

From the Red Hills region to the swamps and mire of the Everglades and from the ti-ti tangles of northern Florida to the mangrove jungles of the southwestern coast the history of surveying in Florida has been one of continuous struggle and perseverance. Cutting through the mangroves proved almost impossible for Albert Gilchrist and a major compromise had to be worked out over an extended period before a proper and adequate resolution could be reached. It was not perfect and could not be so given the relative ignorance of those unfamiliar with the nature of the mangrove swamps. Benjamin Clements lost his son and members of his crew, most of whom were close personal acquaintances, on his survey of the Escambia River country. Sam Hope faced the nearly impossible task of surveying the swamps of Lake Istokpoga and nearly begged to not be required to return to that portion of the state again. Yet these men and others stuck to the job at hand and put out the corners that we rely upon to this very day to identify the property we own.

Robert Butler faced the unenviable task of setting up the first Surveyors General Office in the frontier settlement—one can hardly call it a town at that stage—of Tallahassee. His first message out of this sparsely settled region was by Indian courier. There were few roads and the town was over twenty miles from the coast. Getting materials to run the office, like paper, ink, writing utensils and so forth was difficult at best and sometimes proved impossible. To keep the valuable records safe and dry, Butler had to order the office to be built along specific architectural lines that were difficult to achieve given the lack of available talent in the region. Trying to pay the surveyors when their work was finished was nearly impossible with the local banks discounting the currency of the financial houses upon which the bills were drawn. The rectangular system of surveying was still young enough that the General Land Office [GLO] was willing to experiment with new techniques, like the "compound meander" which made things even more difficult. That Colonel Butler did so well getting talented men to lay out the lines of survey and have them recorded and sold is a solid achievement little appreciated in this day of GPS and other techniques of modern technology.

Benjamin Putnam put up a stiff resistance to the office being used to approve or oversee the implementation of the Swamp and Overflowed Lands Act of 1850. This act put a tremendous burden on his staff and with little direction from the GLO the task demanded much more than he could provide in a timely manner.

The definition of Swamp and Overflowed lands itself left much to interpretation and still plagues the profession to this very day. How wet must the land be to prevent its being used to grow a normal agricultural crop? Where does one draw the line between the river and the neighboring swamp that may or may not be temporarily overflowed during the wet season? Is the line of private verses public property drawn at the open water, swamp line or ordinary high water line and how do you determine each? Such questions made Putnam's job nearly impossible since he did not choose the selecting agents responsible for choosing the swamp lands for Florida. The field notes required by the selecting agents placed an added burden on the limited office staff and in many cases later corrections had to be made before the swamp land selection lists could be forwarded to Washington for final approval. Many of the lands selected were high and dry because only a fraction of the selection had to meet the criteria set up by the act itself. One recent historian/scientist, not understanding the meaning and actual implementation of the act, went so far as to equate the number of acres selected with actual swamp land. The reality, as Benjamin Putnam well knew and predicted, was far different. In fact, Congress passed a clarification of sorts in 1857 which gave all lands then selected, whether swamp or not, to the states. Arkansas wound up getting title to lands in some of the Ozark region high up in the mountains. Such difficulties Putnam saw coming and tried to steer his office clear of the legal and practical problems of the act. In this effort he did not succeed.

These may sound like little technical problems that later surveys and instructions could correct; however, that was not the case. With the state of Florida receiving over 20,000,000 acres under the Swamp and Overflowed Lands Acts these problems continue to plague us today. Surveyors today face the problem of accurately retracing these original lines using the field notes, plats and other available evidence to locate the original corners set up over one hundred and fifty years ago. With the development of the state, the drainage of vast acreages and the changing of river courses by various engineering projects, it is almost impossible to find some evidence of the original corners today. In areas of vast timbering or dredging, the likelihood of finding such evidence is very small. As the trees which were marked as witness trees or mounds built to signify the actual corners have disappeared the task of retracing these corners and the lines upon which they depend has become more and more difficult with each passing year. Because professional standards, the law and frequent court decisions require that a surveyor do such a "retracement," the difficulty in finding the original evidence drives up the costs of such surveys.

The fact that Florida also had a large number of Spanish Land Grants approved by the courts and land commissions also made the surveying of the public lands more difficult. Because most of the lands granted outside of fifteen miles from the city of St. Augustine were in territory held either by the Native American population or contested with intruders from Georgia, these were probably never surveyed by a Spanish surveyor. George J. F. Clark, the man most responsible for surveying lands in the last years of Spanish control, testified that this was the case and that it was simply too dangerous to send any surveyors outside of these limits. Andres Burgevine, one of the more prolific surveyors under Clark's direction, also testified that he was "on the ground but never around it" when he discussed the approved survey of the Great Arredondo Grant. In other words, he did not survey it or put out any corners signifying the boundaries of same. The U. S. Deputy Surveyors who tried to follow in their footsteps found that there were no footsteps to follow or corners to find. For surveyors like F. L. Dancy, this presented a number of problems which often led to charges of fraud and manipulation by those who owned the grants. Following court orders to find surveys never done was an impossible task. It placed a great amount of responsibility on the U. S. Deputy Surveyors to find and lay out the lines of Spanish surveys based upon the best information available, which was often the order itself or a copy of the evidence provided to justify the claim.

Dancy and other surveyors had to match the calls of the Spanish surveys, which were seldom actually done in the field, with the surrounding topography. They also advertised in local newspapers for holders of the grants to provide guidance and information regarding the placement of their grants. Quite frequently this was not provided by the holders of the grants because, in West Florida in particular, the old Spanish law gave proven errors in the surveys to other individuals not the grant holders. The person finding the error usually received some compensation for the discovery and that led to a distrust of U. S. surveyors when they asked for information. Benjamin Clements and James Exum met with this problem constantly in their attempts to survey lands in and around Pensacola. In East Florida where the lands were rarely surveyed in fact the plats of survey provided usually were in error and described lands not found in that location. The discretion of the U. S. surveyors was the only means of ever getting the lands lines done properly. Even here, as in the case of D. A. Spaulding, there were mistakes made.

Spaulding faced the error made by another deputy and a change in the line of survey by a court decision in favor of a powerful political figure. That he succeeded so well in his survey is testimony to his remarkable abilities.

There has been in recent years a rumor spread for political purposes that all of the surveys done in early Florida were unreliable, either through fraud or lack of integrity on the part of the surveyors. One of the most consistent rumors, repeated by an allegedly responsible bureaucrat to two state legislators, is that most of the surveys of early Florida were done by drunken men sitting on bar stools in St. Augustine. This is far from the truth! Only one man was ever accused and proven to match this description and most of his work was quickly disallowed by the Surveyor General of Florida and the GLO. As the essays on John Jackson, Sam Hope, Sam Reid and the others prove, these men were of the highest caliber and trustworthy public servants. Almost all of the early surveyors of Florida were community leaders, highly educated for the day and possessed the trust of their fellow citizens. Over the last fifteen years the author has had the privilege of talking to many of the Public Land Surveyors of Florida today. Many of the author's acquaintances in the field have over fifty years of field experience following the lines of the early surveyors. It is the unanimous feeling among these professionals that given the technology of their day the vast majority of the early surveys of Florida were true representatives of the land surveyed and very accurately done. The men who performed these surveys are held in the highest regard by all concerned and knowledgeable about surveying. Every profession has its "clunkers" and surveying is no exception. However when looking into the history of Florida surveying and the remarkable accuracy of the early work done in a most difficult terrain the results are stunning. Only a hack politician with no knowledge of public lands surveying in Florida would ever spread such defaming rumors about the work of a remarkable group of professionals.

As often noted in seminars conducted by this author, the first surveys of Florida were done with the intent of getting saleable lands on the market. Lands deemed by the surveyor to be too swampy, of poor quality or not suited to cultivation were to be bypassed and picked up as "piece work" at a later date. The sale of public lands was a major source for government revenue at that time and along with the tariff was the largest provider of money to run the government. Getting saleable lands on the market was the highest priority. Contracts for early surveys often called for 600 to 800 miles of lines to be run in a four to six month surveying season. This called for rapid decisions in the field and choices had to be made as to what area to survey and which to leave alone. Because many of the early surveyors and settlers were from northern climates, what today is considered as prime agricultural lands were then looked upon as unsuitable for farming. The rating system, of first, second or third rate lands was derived from areas other than Florida. Hardwood hammocks were considered fertile and most desirable. Sandy pine and palmetto ridge land was looked upon as poor agricultural land and second rate only. Swampy lands covered with cypress islands or wet prairies were considered third rate at best and totally worthless for farming. These were considered as being good for range cattle only, a point picked up nicely in the novels of Patrick Smith. Surveying only in the dry season also meant that some lands were chosen to be surveyed that during the wet season were frequently covered with inches of water and therefore unsuitable for farming. The rush to get the land on the market put tremendous pressure on the Surveyors General and the surveyors of Florida and other states. It also meant that many streams, rivers and lakes were not meandered as provided for in the instructions which also changed often over time. The relatively frequent changes in the instructions also provided for confusion which some have confused with incompetence in surveying. Every survey has to be evaluated as to the instructions under which it is done, the time table allotted for its completion, the type of instrumentation available to the surveyor and many other variables. Simple, snap judgments based upon cursory reviews of the plats or other data will not suffice to determine the accuracy of a particular survey or surveyor.

As for the development of Florida it is easy to blame the Hamilton Disstons of the world for our current ecological problems. Many in the environmental community have rushed to judgment concerning the activities of developers such as Disston, Hope and Gilchrist. However, no one can be considered outside of the time in which they live and the knowledge which we now possess cannot be transmitted backward in time. Our current state of knowledge about the Everglades' importance was not known to Disston or Governor Broward. What they saw was a vast area of rich muck lands perfect for the growth of sugar cane and other such crops. Drain the water from the Everglades and open them for farming was the driving force behind each man, woman and child of that day. Those who succeeded were considered great men. When Disston opened the Cross-Prairie canal to drain Little Lake Tohopekaliga into Big Lake Tohopekaliga it was a big event and opened a vast acreage to farming and development. Kissimmee, St. Cloud and other towns of Osceola County owe their very existence to the drainage conducted by Disston's engineers. Little did anyone know that the drainage of the lakes of the Kissimmee River chain would lead to ecological consequences that we see today. Was the Army Corps of Engineers any better informed when they created the great "Kissimmee Ditch" under great public pressure in the 1950s and 60s? Are we so sure that the restoration of the Everglades as currently planned will actually be better for the environment fifty to one hundred years down the road? Every action man takes to alter the given environment has consequences be it from digging a septic tank, drawing ground water by deep wells or cutting down the surrounding forests. Taken individually the environmental impact of one person's settlement is not drastic, however taken in large numbers and the accumulated impact is great indeed. Finger pointing at one or two developers of the past for current problems is just another way of denying our own responsibilities to the environment.

A study of the life of Hamilton Disston is also interesting from the point of historiography. The rumor of his suicide has been replayed in hundreds of publications many years after the fact by those who never looked into the basis for the rumor. The producer of a television show the author was once involved with took the author to dinner after many protestations about the replaying of the Disston Myth in the script and informed me that the myth would stay in the show even though my evidence was convincing to him. The reason? "It just makes for a better story." Forget the truth it is the story line that matters. Such it has frequently been with our history. When the great diplomatic historian Thomas A. Bailey proved that the myth of "Seward's Folly" was just the editorializing of only one newspaperman, one who happened to be running for president in the upcoming elections of the day, he did not change the interpretation of history still published in nearly every textbook in the United States. It is the story that matters not the fact that this was one of the most unanimously approved decisions ever taken in foreign policy by a sitting president, then under impeachment, is beside the point. To call something this big a "folly" is itself a folly of sorts, yet it is constantly perpetuated by lazy historians and text writers (not always the same people). Primary research is the way to find the actual story but too many are preoccupied with "interpreting" history to undertake this often monotonous task. In history as in the newspaper business, controversy and glitz still sells, the truth be damned.

INDEX

A
Addison, John, 65
Allen, R. C., 7, 140
Alston, Carolina, 64, 74
Alston, Robert, 74
Apalachicola River, 4, 137-138, 140, 145, 147
Apthorp, William Lee, 101-103, 112, 115-116
Armed Occupation Act 1842, ix, 34, 68, 70, 80
Astor, William, 42, 97
Atzeroth, Joseph, 88-89
B
Bacon, Dr. John, 64
Bailey, William, 28, 65
Baldwin, John P., 9
Ball, Leroy, 106-107
Baltzell, Thomas, 29
Barbour, Lt. P.A., 67
Basis Parallel, iii, 5, 7, 136
Battle of Olustee, 41, 60, 157-159
Bemrose, John, 51
Benham, Lt. J., 52
Benjamin, Judah, 162
Benton, Senator Thomas Hart, 31
Big Cypress, 69
Billings, Liberty, 105
Blackburn, Elias E., 28
Blackshear, David, 131
Blake, Thomas H., 74
Bloxham, William, 172-173, 192, 194, 196
Bowlegs, Billy, 84, 87
Braden, Hector, 72
Braden, Joseph, 71
Bronson, Issac, 47, 56
Brown, Thomas, 156
Brush v Prall, 16
Bunker, George, 94
Burch, Lt. Daniel, 131
Butler, Gen. Benjamin, 106
Butler, Robert, iv, 1-8, 32, 56, 80, 99, 132, 139, 147-148, 207
Butler, Thomas, 1-2, 99
C
Call, George W., 17
Call, Richard K., 3, 6, 21-22, 65, 69, 138-139, 156
Caloosahatchee River, 99, 102-103, 115-117, 192, 198, 201
Carter, Gen. Jesse, 151
Casey, Captain John, 35-36, 86
Charlotte Harbor, 66, 71-72, 106-107, 172-173, 175-176, 181, 183
Childs, J. W., 101
Clay, Myron H., 103-104
Clements, Benjamin, v, 7, 119-130, 207, 209
Clinch, Gen. Duncan L., 49-50
Coffee, John, vii, 1, 3, 119-120, 129
Cone, William, 131
Conway, Valentine, 54, 56, 70-74
Cooley, William, 47
Cooper, Gen. T., 67
Corely, Hugh A., 90, 101, 163-164
Crawford, William, 5
Creeks, v, 3, 137
Crystal River, 156, 164
D
Dancy, Francis L., 28, 37, 47-61, 87, 90, 152-153
Daniel, J. J., 94, 200
Darling, John, 43, 47, 89, 155
Levy, David, 57
Denton, James, 25
Dick, John, 58, 94-95
Disston, Hamilton, 185-206
Donalson, J. R., 120-121, 124
Douglas, Thomas, 12
Downing, Charles, 19, 51-52
Drew, George F., 112
Drummund, Judge Josiah H., 111
Drysdale, John, 16
Duval, William Pope, 65-66
E
Ellis, Captain Sam, 107, 181
Emerson, Ralph W., 10
Exum, James, 119, 121, 125-130, 209
F
Fairbanks, George R., 17, 95, 195
Ferguson, George W., 43
Flagler, Henry, 94, 97
Flotard, Thomas, 10
Floyd, Davis, 7
Floyd, John, 131
Floyd, Richard F., 97, 133
Fontane, Mary J., 12
Forbes Island, 138-140
Forbes Purchase, 72, 137, 143-145, 147
Forbes, John, 138, 140, 146
Fort Brooke, 67, 84, 158
Fort Drane, 50-51
Fort Jackson, 120
Fort King, 47, 50
Fort King (Ocala), 47, 50
Fort Marion, 11, 48-49, 51
Fort Meade, 88, 153
G
Gaines, Maj. Gen. E. P., 131
Gamble, James B., 64-65
Gamble, Robert, 63, 65
Gates, Josiah, 67, 70, 72
Gee, John H., 28
Gibbs, James G., 176
Gibbs, Jonathan C., 105
Gilbert, Gen. G. W., 104-105
Gilchrist, Albert W., vii, 107, 169, 171-183, 207, 211
Gleason, William, 43, 111, 165
Goldsborough, Charles, 137, 139-148
Gorrie, John, 9
Great Arredondo Grant, viii, 12, 16-17, 209
Green Swamp, v, 32
Green, Gen. Clay, 2
Greene, Dr. William, 131
H
Hamblin, Samuel, 112, 115-116
Harrison, William H., 2
Hart, I. D., 20
Hartseff [Harsuff], Lt. George, 34, 87
Haulover Canal, 44
Hawkins, George S., 38
Hays, Rachel, 2
Hayward, Col. Robert, 132
Hodgson, R. W. B., 132-136
Hopkins, Benjamin, 57
I
Internal Improvement Fund, 42-45, 57, 112
Irwin, John, xi-xii, 84, 99
Ives, Joseph, 36
J
Jackson, Andrew, iv-v, 1-3, 5, 7, 10, 119
Jackson, Catherine, 79
Jackson, John, vi, 35, 75, 77-78, 80-90, 99-100, 210
Jackson, Rachel, 2
Jacksonville, 151
Jefferson, Davis, 35
Jefferson, Thomas, iv, vii
Jenkins, Horatio, 101, 105-107, 174-175, 181

Jesup, Thomas, 37, 94
Jones, A. H., 33
K
Ker, Robert B., 65, 72-74, 147-148
Key Biscayne, 82
King's Road, 11
Kirby, Ephraim, 11
Kirby, Helen, 11
Kissimmee River, 35, 117, 153, 201, 212
L
Lake Jackson, 5-6, 8
Lake Miccosukee, 75
Lake Okeechobee, 192-193, 198, 201-202
Lanier, Louis, 58
Lanier, Sidney, 60
Lee, Robert E., 160-161
Levy, David [David Levy Yulee], 9, 12, 21, 54, 57, 73, 89, 95, 101
Levy, Moses, 16
Lewis, John H., 17
Lewis, Romeo, vi, 7
Long, Stephen, 48
Lowe, John, 10
M
MacDonald, J. Angus, 112, 114-115
Mackay, George, 82
Madison, ix-x, 24-25, 28-29, 34
Madison County, 29, 32, 94
Maher, Ellen, 81
Mansfield, Captain J.K. F., 52
May, LeRoy, 7
McKay, Alexander, 95
McNeil, 132
McNeil, Daniel, 132-136
McNeill, William, 7
Meade, Gen. George G., 105
Meigs, Josiah, 120
Micanopy, 53
Mickler, J., 94, 165
Miller, A. J., 94-95, 97
Milton, John, 9, 59
Milton, W. H., 177-178
Mitchel, Peter, 12, 16-17
Monroe, James, 4-5, 36
Morris, William W., 84
Moseley, William D., 9, 21
N
Newnansville, 68, 80, 89
O
Ocklawaha River, x, 43, 47, 50
Osteen, John, 28

P
Palatka, 50, 57
Patrick, Lt. M., 66-67
Peace River, 33, 35, 175, 191
Pearce, Charles H., 105
Pensacola, iv-v, 3, 49, 120-121, 123-124, 126, 128, 131, 159, 194, 196
Perry, Madison Starke, 24, 57, 59, 161, 200
Perry, Samuel J., 32
Phillippe, Odette, 47
Phillips, H. H., 20
Pierce, Franklin, 34
Poinsett, Joel, 52
Port Leon, 64-66
Prime Meridian, 5, 7, 119
Putnam, Benjamin, i, vii, 9-25, 33, 47, 56, 83-84, 94-95, 97, 132-135, 207-208
Putnam, Catherine "Kate", 11
R
Randall, Thomas, 16-17
Randolph, Arthur M., vii-ix, 7, 65, 95-96, 113, 132, 135-136
Reed, Gov. Harrison, 105, 111
Reid, Carolina, 74-75
Reid, Florida, 49
Reid, Robert Raymond, 49, 51
Reid, Sam [Samuel], 63-75, 80, 105, 210
Rodman, John, 16
Russell, Charles, 31
S
Scarlet, R. L., 180-183
Scott, Gen. Winfield, 49
Seawell, Captain William, 67
Seminoles, iv, vi, 3-4, 15, 36, 50, 53, 84, 86-87, 114, 137
Smith, C. F., 107, 113, 116
Smith, Edmund Kirby, 10
Smith, Joseph L., 10, 16
Southern Life Insurance and Trust Company, 14
Spaulding, 97
Spaulding, D. A., 93-97
St. Augustine, i, iv, x, 5, 31, 33, 37, 39-43, 47-53, 59, 61, 75, 81-82, 84, 93-94, 97, 120, 131, 209-210
St. Johns and Indian River Canal Company, 40, 43
St. Johns Railroad, 43, 97
St. Joseph, 21
St. Joseph's Convention, 55
St. Marks, iii, 4, 141

St. Marks Fort, 139-140
St. Marks Railroad, 39
St. Marks Reserve, 146
St. Marks River, 138, 145
St. Marks, Port of, 66
Stearns, Marcellus, 100-103, 106, 109-114, 118
Swamp and Overflowed Lands, ix, 34, 40, 42, 94, 190-191, 196, 207-208
Swann, Samuel, 101
T
Tacony, 187-189
Tallahassee, iii, viii, 5-7, 29, 34, 37, 45, 49, 54, 63-64, 70, 72, 74, 83, 90, 114, 132, 163, 168, 181, 194, 207
Tallahassee Rail Road, 64-65
Tampa, 35-36, 41, 67-68, 71, 77, 80-81, 83, 89, 150, 153, 155, 159, 167, 172, 195
Tampa Bay, 66-67, 80, 89
Tannehill, James, 101
Tanner, John, 9
Thompson, Wiley, 131
Trabue, Issac, 172
Turman, Simon, 80, 89, 155
U
Union Bank of Florida, 14, 18
V
Vail, J., 64, 66
Vose, Francis, 43, 57, 112, 190-191, 193
W
Walker, David, 64-66
Walker, George K., 66
Wanton, Edward, 10
Washington, Henry, vii-viii, 7, 68-69, 71
Watson, Col. James C., 132
Westcott, James D., ix
Westcott, John, i, vi, viii-x, 27-45, 55, 58, 84-85, 88, 93-97, 116, 166-167
Whidden, John W., 176
Whitcomb, James, 144
White, Thomas, 7, 125, 128-129
Whiting, Maj. L., 83
Whitner, Benjamin, vii, 7, 132
Williams, Marcellus, 101, 106, 113, 115, 117
Wilson, John, 35, 95
Withlacoochee River, 32, 50
Worth, William J., 23, 66-68, 86, 117